T0235244

SpringerBriefs in Molecular Science

Biobased Polymers

Series editor

Patrick Navard, CNRS/Mines ParisTech, Sophia Antipolis, France

Published under the auspices of EPNOE*Springerbriefs in Biobased polymers covers all aspects of biobased polymer science, from the basis of this field starting from the living species in which they are synthetized (such as genetics, agronomy, plant biology) to the many applications they are used in (such as food, feed, engineering, construction, health, …) through to isolation and characterization, biosynthesis, biodegradation, chemical modifications, physical, chemical, mechanical and structural characterizations or biomimetic applications. All biobased polymers in all application sectors are welcome, either those produced in living species (like polysaccharides, proteins, lignin, …) or those that are rebuilt by chemists as in the case of many bioplastics.

Under the editorship of Patrick Navard and a panel of experts, the series will include contributions from many of the world's most authoritative biobased polymer scientists and professionals. Readers will gain an understanding of how given biobased polymers are made and what they can be used for. They will also be able to widen their knowledge and find new opportunities due to the multidisciplinary contributions.

This series is aimed at advanced undergraduates, academic and industrial researchers and professionals studying or using biobased polymers. Each brief will bear a general introduction enabling any reader to understand its topic.

*EPNOE The European Polysaccharide Network of Excellence (www.epnoe.eu) is a research and education network connecting academic, research institutions and companies focusing on polysaccharides and polysaccharide-related research and business.

More information about this series at http://www.springer.com/series/15056

Stefan Spirk

Polysaccharides as Battery Components

Springer

Stefan Spirk
Institute for Paper, Pulp and Fiber
    Technology and Institute for Chemistry
    and Technology of Materials
Graz University of Technology
Graz
Austria

ISSN 2191-5407   ISSN 2191-5415   (electronic)
SpringerBriefs in Molecular Science
ISSN 2510-3407   ISSN 2510-3415   (electronic)
Biobased Polymers
ISBN 978-3-319-65968-8   ISBN 978-3-319-65969-5   (eBook)
https://doi.org/10.1007/978-3-319-65969-5

Library of Congress Control Number: 2017951153

Printed on acid-free paper

This Springer imprint is published by Springer Nature
The registered company is Springer International Publishing AG
The registered company address is: Gewerbestrasse 11, 6330 Cham, Switzerland

Contents

1 **Introduction** 1
  1.1 A Brief History on Batteries—Evolution of New
      Technologies 2
  1.2 Basic Considerations for Battery Components 5
      1.2.1 Safety Aspects................................... 5
      1.2.2 Material Aspects 6
  References 8

2 **Polysaccharides in Batteries** 9
  2.1 Polysaccharides as Binders in Batteries 9
      2.1.1 Comprehensive Data on Different Electrode
            Materials Using CMC and Other Polysaccharides 16
  2.2 Polysaccharides as Separators 30
      2.2.1 Microporous Membranes 31
      2.2.2 Composite Membranes........................... 34
      2.2.3 Non-woven Mats................................ 36
      2.2.4 Solid Polymer Electrolytes (SPE), Gel Polymer
            Electrolytes (GPE) and Composite Polymer
            Electrolytes 39
  2.3 Electrode Materials from Polysaccharides.................. 49
  References 53

3 **Conclusion and Outlook** 59

Abbreviations

AA	Agar-agar
AQ	9,10-anthracenedione
β-CD$_p$	Polymerized branched cyclodextrin
BC	Bacterial cellulose
BMIM	1-Butyl-3-methylimidazolium
CA	Cellulose acetate
CAB	Cellulose acetate butyrate
CBC	Carbonized bacterial cellulose
C-CNF	Carbonized cellulose nanofiber
ChNC	Chitin nanocrystal
CMC	Carboxymethyl cellulose
CMCh	Carboxymethyl chitosan
CN-CMCh	Cyanoethylcarboxymethyl chitosan
CNC	Cellulose nanocrystal
CNF	Cellulose nanofibers
CN-HPC	Cyanohydroxypropyl cellulose
CNT	Carbon nanotube
CV	Cyclovoltammetry
DMC	Dimethylcarbonate
DSSC	Dye sensitized solar cell
EC	Ethyl carbonate
EMIM	1-Ethyl-3-methylimidazolium
FEC	Monofluoro ethylene carbonate
FEG	Field emission gun
GA	Gum arabic
GG	Guar gum
GO	Graphene oxide
GPE	Gel polymer electrolyte
HEC	Hydroxyethyl cellulose
HPC	Hydroxypropyl cellulose

ITO	Indium tin oxide
LFP	Lithium iron phosphate, $LiFePO_4$
LIB	Lithium ion battery
LiBOB	Lithium bis(oxalato)borate
LTO	Lithium titanate, $Li_4Ti_5O_{12}$
MCC	Microcrystalline cellulose
MCMB	mesocarbon microbeads
MHII	1-methyl-3-hexylimidazolium iodide
MPII	1-methyl-3-propylimidazolium iodide
NMC	Nickel-manganese cobalt oxide, $LiNi_{0.33}Mn_{0.33}Co_{0.33}O_2$
NMP	N-methyl-2-pyrrolidone
P3HT	Poly-3-hydroxythiophene
PAA	Polyacrylic acid
PANI	Polyaniline
PE	Polyethylene
PEDOT	Poly(3,4-ethylenedioxythiophene)
PEG	Polyethylene glycol
PET	Polyethylene terephthalate
PLLA	Poly-L-lactic acid
PMMA	Polymethylmethacrylate
PP	Polypropylene
PPY	Polypyrrol
PSS	Polystyrene sulfonate
PVA	Polyvinyl alcohol
PVDF	Polyvinylidene fluoride
PVDF-HFP	Polyvinylidene fluoride-co-hexafluoropropylene
rGO	Reduced graphene oxide
SBR	Styrene-butadiene rubber
SIB	Sodium ion battery
SPE	Solid polymer electrolyte
XG	Xanthan gum

Abstract

This work aims to provide a broad overview on the use of polysaccharides in battery applications and is divided into three sections. The first section introduces the reader into the historic development of batteries and the evolution of different battery chemistries and design. In this chapter also the different prerequisites for battery components will be briefly discussed. In the second part, battery binders are in the main focus, whereas lithium ion batteries are among the most prominent congeners. The third chapter deals with polysaccharides in separators, solid state electrolytes, polymer gel electrolytes, and composite gel electrolytes. The last section deals with so called bio-carbons, prepared by carbonization of polysaccharides to create electrode materials for battery applications.

Basic Parameters and Definitions for Battery Performance

Capacity   Amount of electric charge a battery can deliver at the rated voltage (A·h)

C-rate   Discharge current divided by the theoretical current draw under which the battery would deliver its nominal rated capacity in one hour

Rated Capacity   Capacity of the battery expressed as the product of 20 hours multiplied by the current that a new battery can consistently supply at room temperature

Specific capacity   Amount of electric charge a battery can deliver at the rated voltage per the mass of active electrode material ($A·h·g^{-1}$)

Chapter 1
Introduction

In the past decades, environmental concerns have been urging mankind to rethink its energy supply strategy from non-renewable fossil sources towards sustainable and renewable ones. Wind, solar, and tidal energy are intermittent in nature; however, the generated electricity usually does not match demand. As a consequence, efficient electrical energy storage is a key technology to shift energy supply from fossil based to renewable energy. Further, dependable and high electrical energy storage solutions must be developed for the major source of greenhouse gas emissions, namely traffic. Electrochemical energy storage is capable to overcome these issues and in the case of transport it is an essential part to solve the greenhouse problem. Major storage technologies are supercapacitors and rechargeable batteries. Supercapacitors are already used in many everyday products including mobile devices, load shaving, and electrically operated trains. Batteries have been revolutionizing portable electronics and are now being pushed into electric vehicles and stationary storage markets. However, to realize a replacement of current technologies and to achieve a significant economic and environmental impact, several challenges must be overcome for both storage technologies. To address this challenge, a new generation of devices with higher energy density, lower cost, longer lifetime and improved materials sustainability is required.

In this book, we will mainly focus on polysaccharides and batteries since an overview on supercapacitors has been described in a separate report [1]. The major difference between batteries and supercapacitors is the electric storage mechanism. In batteries, redox reactions in bulk electrode materials are of utmost importance whereas in supercapacitors the surface is decisive; therefore large surface areas are required for electrode materials for supercapacitors which is not a major issue in batteries.

We shall first examine the working principles and characteristics of batteries and will focus on the historical evolution of the different battery technologies. In the main chapters, we will discuss the requirements of battery components and how polysaccharides are capable to meet these. The major applications described in this book comprise binders for electrode materials for different type of battery chemistries, and separators in all their forms such as different types of membranes,

© The Author(s) 2018
S. Spirk, *Polysaccharides as Battery Components*, Biobased Polymers,
https://doi.org/10.1007/978-3-319-65969-5_1

solid state electrolytes, polymer gel electrolytes and composite gel electrolytes. In the end, carbonaceous materials made from polysaccharide for battery applications complement the book.

1.1 A Brief History on Batteries—Evolution of New Technologies

Like many discoveries in science and technology, the invention of batteries was a coincidence. In 1780, it was Luigi Galvani of Italy, who recognized that during dissecting a frog, its legs twitched when they came in contact with the iron scalpel. He believed this phenomenon to be caused by 'animal electricity' [2]. However, his friend and fellow Alessandro Volta strongly disagreed and suspected the metal surfaces joint by a moist environment to be responsible for Galvani's observations. Finally, Volta proved his hypothesis and his investigations finally led to the development of the first battery, the so called voltaic pile. This battery was composed of piled copper and zinc disks which were separated by cloth or cardboard to avoid short circuits [3]. These first, cellulose based separators were pre-soaked in an electrolyte solution. This battery enabled a stable supply of electricity and current while exhibiting hardly any self-discharge. However, the original setup featured some drawbacks such as electrolyte leakages due to the weight of the stack, as well as short battery lifetime. While the first problem was simply solved by arranging the stack in a vertical way (a so called trough), the second problem was more difficult to tackle. The origins of the low battery lifetime were based on the evolution of hydrogen, steadily increasing the internal resistance of the copper electrode, as well as the degradation of the zinc anode caused by impurities deposited during operation, finally leading to short circuits. The corrosion of the zinc anode was avoided by the use of amalgated zinc preventing the formation of impurities at the electrode. The issues with hydrogen evolution led to the design of a new cell type, the Daniell cell. This cell was composed of a zinc anode, and a copper vessel filled with copper sulfate, in which a porous, ceramic tray filled with sulfuric acid was placed into. The ceramic layer allowed for an exchange of ions but prevented mixing of the solutions. The Daniell cell was the first commercial battery type having a voltage of 1.1 V with a much longer life time than the Voltaic cells since instead of insulating hydrogen, conducting copper metal is deposited on the anode over time [4]. The importance of the Daniell cell in battery history is reflected in the unit of the electromotive force, Volt, since the operating voltage of this cell was ca 1 V [5].

Later on, some improvements of the Daniell cell were introduced such as the Bird cell (here gypsum plaster was used as porous barrier) [6], the Gravity cell and the Poggendorf cell [7]. Particularly, the Gravity cell was a major step forward in battery technology since it avoided the use of a barrier, thereby significantly reducing the internal resistance while increasing battery current. This cell consisted of a glass jar whereas the copper electrode was located on the bottom of the cell, and a zinc electrode on the top of the jar. The jar was then filled with copper sulfate crystals deposited around the copper electrode. After addition of water, a copper sulfate solution is formed and as soon as a current is drawn, a top layer of zinc sulfate is formed around the zinc electrode which remains separated from the copper sulfate solution by its lower density. This type of battery was widely used in British and American telegraph networks until the mid 1950s.

The Leclanche cell was developed in 1866 and used ammonium chloride as electrolyte, manganese(IV) oxide with some carbon as cathode materials, and zinc as anode. The addition of carbon to the cathode material increased the conductivity and absorption of ions due to the higher surface area [8]. The cell provided 1.4 V.

In contrast, the Poggendorf cell employed a different concept [7]. As electrolyte, a mixture of diluted sulfuric and chromic acid was employed and the electrodes were composed of carbon (cathode) and zinc (anode). The cell provided 1.9 V which was rather high for a battery in the 19th century.

The introduction of sulfuric acid as battery electrolyte paved the way for the development of a new generation of batteries which were potentially rechargeable. Until this point, batteries' lifetime was limited by the amount of redox active compounds in the battery (so called primary batteries). The most prominent example included the lead acid battery. In such a battery, the anode material consists of elemental lead while the cathode material was made of lead dioxide. As electrolyte, sulfuric acid is used which reacts with both electrodes to form lead sulfate. However, while at one electrode electrons are produced by the formation of the sulfate at the other one they are released, thereby creating a current. Since the process is reversible, the battery can be recharged by reversing the current.

Different improvements have been made to increase the performance of the battery such as the Planté cell, which used lead sheets separated by rubber and wound into a spiral. The principle of a lead battery has not changed since its invention and still nowadays these batteries are in use for applications where weight is not a crucial factor such as automotive industry for instance.

A completely different approach was described in the beginning of the 20th century which was based on alkaline electrolytes, mostly potassium hydroxide, in combination with cadmium and nickel electrodes. These batteries had a higher energy density but were significantly more expensive than the lead acid type batteries.

The next major step in battery development was the invention of so called dry cells. These cells use paste like electrolytes which contain just enough moisture to allow for a current flow. One of the major advantages is that these cells can be used in any direction since they do not contain any electrolyte which could potentially be spilled off. This opened the door for portable applications of batteries. The zinc carbon battery was the first of its kind and was an improved version of the Leclanche cell. The main difference was that instead of a liquid electrolyte, a paste made from plaster of Paris or wheat flour was used, which contained ammonium or zinc chloride. Later on, the design of the cell was altered and nowadays the cell (1.5 V) is composed of a zinc casing serving at the same time as anode, a manganese(IV) oxide cathode (which contains powdered carbon) which is attached to a carbon rod, which is conductive and inert against corrosion by the electrolyte. Nowadays, the most used gel electrolytes consist of a moist paste of ammonium or zinc chloride impregnated on paper, which serves as separator between the zinc anode and the manganese(IV) oxide cathode. Nowadays, this type of cell accounts for a large share (ca. 20%) of the market for portable batteries.

Like zinc carbon also alkaline batteries have been developed into dry cells, which significantly boosted their use. As a gel electrolyte, potassium hydroxide is used while as anode and cathode materials zinc and carbon are implemented. The separator consists of a nonwoven fabric, often based on cellulose. The cell features a voltage of 1.4 V and nowadays is one of the most commonly used primary batteries with market shares of ca 50%.

The latest stage of evolution in batteries refers to the need for higher voltages of batteries associated with the need for higher energy density while being rechargeable. Originally invented in the beginning of the 20th century by G.N. Lewis, it took another 60 years until this technology became exploitable, particularly due to the efforts of Besenhard and later Goodenough [9]. A scheme of the design of such batteries is depicted in Fig. 1.1.

The first commercialized lithium ion battery (LIB) was brought to the market by Sony in 1991. LIBs can deliver voltages of over 3 V. Therefore, electrolytes used in LIBs must be stable at these conditions; water does not meet this requirement (decomposition at 2 V). Commonly used electrolytes in LIBs are ethylene carbonate or diethyl carbonates which contain lithium complexes with non-coordinating anions such as $LiPF_6$ or $LiBF_4$. As cathode material, graphite is

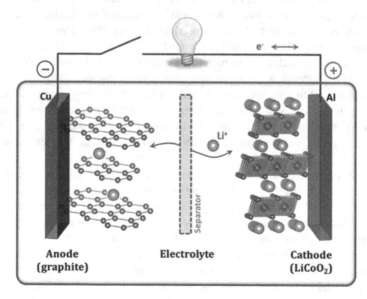

Fig. 1.1 Schematics of a lithium ion battery using $LiCoO_2$ and graphite as electrode materials. Reprinted with permission from the American Chemical Society [10]. Copyright (2013) American Chemical Society

very often commercially used while as anode three types are extensively studied, namely layered oxides, polyanions or spinels. During charge and discharge cycles, lithium ions are moved from one electrode to the other where they are intercalated/removed. Nowadays, LIBs are the most commonly used battery type to power consumer electronics but lately also other areas turned into the focus such as electrically powered mobility solutions.

Lately also other battery types are being developed since there will be probably problems in terms of availability of raw materials (i.e. lithium supplies) to come. Such technologies avoid the use of lithium in their active electrode materials and aim at incorporating sodium magnesium or other metals for the generation of high capacity batteries.

1.2   Basic Considerations for Battery Components

As we already discussed in the previous section, different aspects in the design of batteries must be considered. Some of these aspects are based on safety, others relate to the performance of the cell while others consider economic factors, which are not explicitly discussed here. However, all these factors go hand in hand and must be considered when designing new battery systems. In the following, an overview is given on the most important factors.

1.2.1   Safety Aspects [11]

This issue can be divided into several subcategories: Cell chemistry, electrolytes, cell design, and separator failure. The cell chemistry is an important factor when it comes to the most important parameter of batteries, the energy density. High energy density is always associated with higher chemical reactivity which increases the risk of undesired side reactions in a battery.

Therefore in commercial applications, the use of systems with lower energy density may be preferred over high energy systems when safety is an important issue. Lithium cobalt oxides for example, which provide high energy densities are rather reactive and have been replaced by other less reactive chemistries (e.g. lithium phosphates) in order to overcome safety issues in LIBs. The electrolytes in turn must be very often stabilized in order to prevent decomposition/degradation which is particularly favored at high cell voltages. Cell design is usually not an issue for the traditional types of batteries having rather low energy density. However, for LIBs for instance this can be a problem, particularly if heat transport has not been appropriately considered during cell design. If this is not the case, the cell may suffer from local hot spots which may lead to cell failure. Overheating of the cell for whatever reason is a potential threat to separators, which are usually made from polymers. A very prominent example for cell design failure is the Galaxy S7 produced by Samsung, which has been retracted from the market by the end of 2016. If the cells undergo a thermal runaway, the separators may melt and may cause a short circuit which may lead to severe fires. However, there are a variety of special separator designs in order to circumvent these cases such as thermally stable rigid separators.

1.2.2   Material Aspects [12]

Besides safety, the choice of materials is crucial to determine the performance of cells. The most important aspect is the cell chemistry in a galvanic cell. It defines the nominal voltage and is a major criterion in the applicability of a battery in practice. However, also the shelf life is dependent on the cell chemistry; Leclanche cells typically last for 2–3 years, while alkaline batteries can be used up to 5 years while LIBs may be in use for 10 years. Similar issues relate to self-discharge, which happens when the battery is not in use (or stored in a shelf) and a current is not drawn from the battery. This self-discharge strongly varies between the different types of cell chemistries and range from 2–3% per month for LIBs to 4–5% for lead acid batteries and up to 15–20% for NiCd types. Another important factor promoting self-discharge is temperature. At elevated temperatures, undesired side reactions are more favored and as a consequence self-discharge proceeds faster and shelf life is reduced. Besides these shelf storage considerations, other parameters are important for the use of batteries. This includes discharge rates, cycle life and surface area of the materials. Discharge rates influence the battery capacity, i.e. at very high discharge rates the capacity is reduced and vice versa. Cycle life refers to as the lifetime of the battery since it is a direct measure to evaluate the battery capacitance in dependence of charge/discharge cycles. Usually, cycle life is defined by the number of charge/discharge cycles a cell can perform until its capacity reaches 80% of its initial capacitance. A similar parameter is the Total Energy Throughput. It relates to the amount of energy in Wh which can be drawn from a battery until 80% (during cycling) of its initial capacity are reached. The surface area of the material is a crucial point for the judgement of battery materials since capacitance, energy and power are usually normalized by the weight or volume. In this context, values per mass of the electron conductor are often provided, which may, albeit technically correct, be highly misleading with regard to "true" values one can expect at the device level. This habit to correlate performance metrics of electrochemical energy storage devices to the mass or volume of a certain "active" component has been become common for both batteries and supercapacitors [13]. However, for both applications, supercapacitors and so called beyond-intercalation batteries, the reported electrochemical performance parameters may represent just a part or even a negligible fraction of the total device mass or volume. While this practice would be acceptable if all experimental details have been made available in scientific papers, very often in publications information to derive values for full electrode or device parameters including all other necessary components are not provided. This problem has been recognized for both supercapacitors and batteries [14], but best-practice publishing standards as recently proposed for solar cells are, however, yet to be defined [15]. A material may show exceptional performance when values are referenced to the material alone, but inclusion of more cell components may alter these values massively. Particularly nanomaterials such as graphene have a very low packing density. The empty space has to be filled with a large mass of electrolyte without adding capacity.

A very important plot to showcase the different characteristics of the different storage systems is the Ragone plot (Fig. 1.2). This type of plot provides the typical ranges of specific power and specific energy for capacitors and batteries in double logarithmic scale.

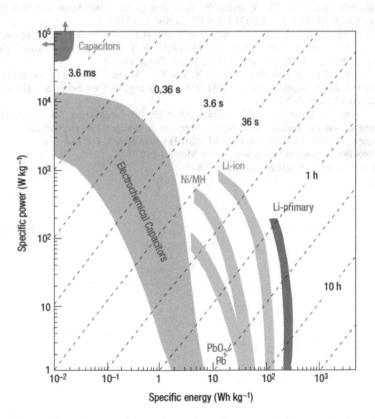

Fig. 1.2 Ragone plot with accessible regions of specific power and energy for batteries and electrochemical capacitors. Times associated with the diagonal lines are time constants obtained by dividing energy by power. Reproduced from Ref. [16] with permission from Nature Publishing Group

References

1. Liew SY, Thielemans W, Freunberger S, Spirk S (2017) Polysaccharide based supercapacitors. Springer, Berlin
2. Galvani L, Volta A (1791) De viribus electricitatis in motu musculari commentarius retrieved from https://archive.org/details/AloysiiGalvaniD00Galv
3. Dibner B (1964) Alessandro Volta and the electric battery. Franklin Watts
4. Spencer JN, Bodner GM, Rickard LH (2010) Chemistry: structure and dynamics. Wiley, New York
5. Hamer WJ (1965) National bureau of standards monograph #84. US National Bureau of Standards
6. Bird G (1838) Report of the seventh meeting of the british society for the advancement of science, London, retrieved at http://www.biodiversitylibrary.org/item/46624#page/5/mode/1up
7. Ayrton WE, Mather T (1911) Practical electricity. Cassell and Company, London
8. Saunder SA (1875) Nature 12:564
9. (a) Besenhard JO, Eichinger G (1976) J Electroanal Chem Interfacial Electrochem 68:1; (b) Eichinger G, Besenhard JO (1976) J Electroanal Chem Interfacial Electrochem 72:1; (c) Mizushima K, Jones PC, Wiseman PJ, Goodenough JB (1980) Mater Res Bull 15:783
10. Goodenough JB, Park K-S (2013) J Am Chem Soc 135:1167
11. (a) Rezvanizaniani SM, Liu Z, Chen Y, Lee J (2014) J Power Sour 256:110; (b) Abada S, Marlair G, Lecocq A, Petit M, Sauvant-Moynot V, Huet F (2016) J Power Sour 306:178
12. Nitta N, Wu F, Lee JT, Yushin G (2015) Mater Today 18:252
13. (a) Choi N-S, Chen Z, Freunberger SA, Ji X, Sun Y-K, Amine K, Yushin G, Nazar LF, Cho J, Bruce PG (2012) Angew Chem Int Ed 51:9994; (b) Bruce PG, Freunberger SA, Hardwick LJ, Tarascon J-M (2012) Nat Mater 11:19
14. (a) Gogotsi Y, Simon P (2011) Science 334:917; (b) Freunberger SA (2015) In: ECS conference on electrochemical energy conversion & storage; The electrochemical society: Glasgow; (c) Obrovac MN, Chevrier VL (2014) Chem Rev 114:11444
15. A checklist for photovoltaic research. Nat Mater 14:1073
16. Simon P, Gogotsi Y (2008) Nat Mater 7:845

Chapter 2
Polysaccharides in Batteries

As already briefly indicated in the first sections of this book, polysaccharides and here particularly cellulose, have been used from the very beginning of the battery age. In principle, either the native polysaccharide or modifications thereof can be used for battery applications. Alternatively, polysaccharides can be converted into carbonaceous materials.

As already discussed above, polysaccharides have been integral parts of early battery designs and still today they are wide-spread in different applications on the market. While in the early stages of battery design, cellulose in the form of paper, board and cloth was used as separator, in recent years applications extended to binders, solid state electrolytes and as precursor for carbon electrodes.

2.1 Polysaccharides as Binders in Batteries

The binder is a crucial component of batteries, particularly in LIBs where energy density is a major issue. The role of the binder is to protect the electrode material against the electrolyte, while allowing for ion migration throughout the binder. Therefore, huge efforts have been made to improve the performance of binders for nanosized active electrode materials such as silicon or tin alloy nanoparticles. These nanoparticles undergo huge volume changes upon lithiation/delithiation (200–300%), which leads to mechanical stress of the binder (Fig. 2.1). After a certain amount of charge/discharge cycles, the binder cannot accomplish any more for the volume changes due to lithium insertion/removal, which results in the formation of cracks due to mechanical fatigue. As a consequence, the electrolyte can access the active electrode material, leading to pulverization or graining concomitant with a loss in performance due to non-contacting. At a certain point, the battery is not working any more, since the grains are too small.

A widely used commercial binder is polyvinylidene fluoride (PVDF), which is used for both anode and cathode materials. Its use originates from its rather high

© The Author(s) 2018
S. Spirk, *Polysaccharides as Battery Components*, Biobased Polymers,
https://doi.org/10.1007/978-3-319-65969-5_2

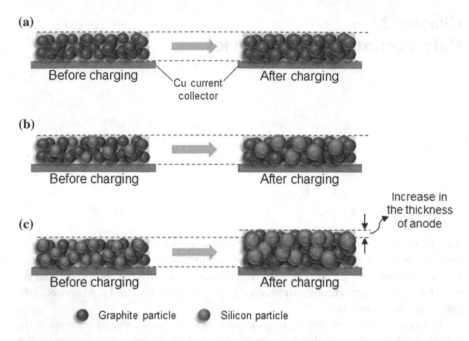

Fig. 2.1 Schematic showing cross-section of anode before and after charging for **a** graphite based anode, **b** SCC anode with low amount of Si, and **c** SCC anode with high amount of Si. As can be seen, the anode thickness remains unchanged for graphite and low Si based SCC anode. Swelling of anode is experienced for anode with higher Si content (**c**). Reproduced from Scientific Reports using a Creative Common License 4.0 from Ref. [1]

electrochemical stability, as well as its rather good adhesion and compatibility with electrode materials and current collectors. However, the major drawback of PVDF is its rigidity, which creates problems to account for volume changes during charging/discharging cycles. In addition, binder concentration should be kept to the minimum to reach high energy densities.

Therefore, alternative approaches have been developed using polysaccharides as major components in the formulation of binders. Like PVDF, there are many polysaccharides which readily form homogeneous films and layers on a wide range of materials. In addition, many polysaccharides (and derivatives) are water soluble and easy to process since organic solvents can be avoided during cell manufacturing. The oldest polysaccharide derivative used as binder is carboxymethyl cellulose (CMC).

The use of CMC as binder is nowadays widespread and can be considered state of the art, besides PVDF. Initially, it was believed that CMC as a 'soft' material is able to accommodate volume changes by its 'soft' nature. However, this is not exactly true, since CMC is a rather stiff material which features a high degree of brittleness having just 5–8% deformation at break. Already some time ago, it was shown that the concept of 'soft' elastomeric materials for binders is just a part of the story. This example involved the comparison of a SBR grafted CMC and a pure CMC binder. Interestingly, the stiffer CMC binder performed much better than the elastomeric SBR-CMC one in all relevant electrochemical performance parameters [2].

Regarding the envisaged volume changes during lithiation of nanosized silicon powders, the mechanical stress induced by this change should lead to failure in the case of brittle CMC. Therefore, investigations went into different directions to evaluate the reasons for better performance. One finding was that CMC favors the dispersability and crosslinking of conductive additives such as carbon black and the nanosized silicon powders during processing. CMC is soluble in water due to dissociation of the COOH groups. These groups form in concentrated particle/polymer suspensions such as electrode slurries, a three-dimensional network by crosslinking the particles to polymer chains. This originates from adsorption of different segments of individual CMC macromolecules on different particles thereby forming entanglements (Fig. 2.2). By choosing the right drying conditions for removal of the water, this three dimensional network retains their morphology and structure in dry state. The efficiency of physical crosslinking is influenced by the conformation (coiled vs. extended) and the molecular mass of the polymer. Coiled conformations in combination with high molar mass of the polymer facilitate the formation of crosslinks, thereby improving the performance of potential binders. It was shown that CMC can accomplish for ca. 400% in volume change without a loss in performance [3].

Besides these physical interactions as proposed by Li and Lestriesz, also chemical bonds between the carboxylic groups of CMC and the surface of the Si-NPs can be formed, depending on the pH value used for slurry preparation. These ester type bonds are preferentially formed by an esterification of surface bound Si–OH silanol groups with the deprotonated carboxy group which was proven by ATR-IR and ^{13}C solid state NMR spectroscopy [5]. The reason for the importance of the pH value is based on the pKa values of Si–OH and COOH

Fig. 2.2 Two different binding principles that can be used for preparation of composite electrodes consisting of powders: **a** direct binding by adsorption of macromolecules on neighboring particles and forming interparticle bridges. **b** Indirect binding by forming a 3D network into which particles are mechanically entrapped. Reproduced with permission from Elsevier from Ref. [4]

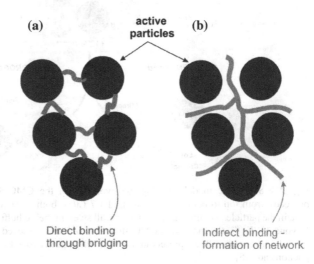

(a)   active particles   (b)

Direct binding through bridging

Indirect binding – formation of network

groups. At pH 3 using a buffered solution, both groups are protonated which favors condensation reactions in the slurry during drying, which has been shown extensively by Mazouzi et al. [6]. Using this knowledge, over 700 cycles have been realized by reducing the amount of stress on the binder by reduction of Li-insertion into the active material by ca. one third (capacity 960 mAh g^{-1}). Similar observations (capacity 1200 mAh g^{-1}) have been made by Tranchot et al. [7] who investigated micrometric Si particles. Again, electrodes prepared at pH 3 performed much better than those prepared at pH 7. Lower irreversible expansion was observed for the pH 3 electrodes at the end of the 1st cycle ($\sim$50% compared to $\sim$180% for the pH7 electrode).

However, the presence of a defined oxide layer on the silicon nanopowders is crucial for the realization of a good binder system as shown by Delpuech [8]. Consequently, a lack of a defined oxide layer leads to reduced cycle life, even when the slurries have been prepared at pH 3 using buffered solution.

Bridel proved that the degree of polymerization is a crucial factor for a good cyclability. Long chains of macromolecules feature a higher probability to create bridges between COOH and Si–OH groups. As a consequence, in the case of failure, this may lead to self-healing properties of the Si–OH–COOH couple [9]. Further, in situ SEM investigations showed that the Si/CMC/carbon material can accomplish for the volume changes (expansion and shrinking) during charge/discharge. They also showed that the volume change occurs in two steps. At Li:Si ratios in the range of 1.7–2:1, the lithium can be accommodated in the pores of the electrode material. At higher Li:Si ratios, the electrode material rearranges and the CMC-Si network changes its structure by altering the hydrogen bonding mode

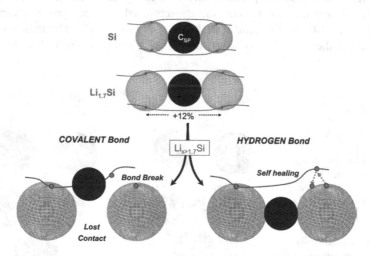

Fig. 2.3 Schematic model showing the evolution in the CMC-Si bonding as the Li uptake proceeds, from top to bottom. Up to around 1.7 Li/Si, both covalent and hydrogen bonding can sustain the particles volume changes, the overall swelling being buffered by the electrode porosity. Beyond 1.7 Li/Si, the maximum CMC stretching ability is reached, and only the hydrogen-type Si-CMC interaction allows preservation of the efficient network through a proposed self-healing phenomenon [5]

(Fig. 2.3). As a consequence, the battery retains its reversible capacity by self-healing as mentioned above.

The interaction of lithium ions and CMC was investigated as well by means of variable-time pulsed field gradient spin-echo (PFGSE) NMR spectroscopy [10]. The major result of that study was that even at concentrations where Li-CMC had a gel-like behavior, Li^+ diffusion coefficients are very close to that in water. As expected, the Li^+ ions mainly interact with the COOH groups of CMC, thereby providing a rather complex coordination pattern. Nevertheless, as shown by these studies, the diffusion of Li^+ in these systems is essentially unrestricted, with a fast, nanosecond-scale exchange of the ions between CMC and the aqueous environment.

The interaction at the nanometer level was studied by Maver et al. [11] using AFM at different pH values used for the preparation of electrodes. Their results showed that there is a relationship between the measured interaction forces between the CMC and Si and the mechanical strength of the electrode materials. Further, this study revealed that covalent CMC-Si bonds feature forces which are ca. one order of magnitude larger than those of physically bound CMC on Si. Simulations using a coarse grain model corroborate these findings [12]. In such a model, the macromolecules are assembled into beads with a size of their corresponding Kuhn-length. Despite being simple, this model yielded quantitative predictions on CMC adsorption on Si, which were in good agreement with experimental results in several respects. The model revealed the same mechanism as suggested above (bridging of Si-NPs by hydrogen and covalent bonds) and also provided possible explanations why volume change can be accommodated by CMC. It was shown that CMC molecules are arranged in a spring like conformation. This conformation is very flexible and upon volume changes, the spring gets stretched and maintains bridges between the Si particles.

The second mechanism leading to capacity fading of the Si anodes is the degradation of the electrolyte. The degradation of the electrolyte induces the growth of an SEI layer on the active electrode materials, thereby blocking pores and impairing diffusion of lithium ions into the anode [13]. Some studies showed that CMC may already act as a kind of artificial SEI layer on composite electrode materials since the loss of capacity after the 1st cycle is 10–15 times lower compared to classical PVDF binders. Further, it was proven by XPS that even partial covalent crosslinking of the CMC with Si-NPs led to a significant reduction of $LiPF_6$ degradation [8, 14]. Jeschull et al. [14] highlighted the importance of the interaction of different conductive additives having different surface areas with the binder and the active electrode materials. It turned out that higher specific surface additives tend to induce more cracks than those with lower surface area and feature a much lower cycle life. Further, the tendency for cracks was influenced by the type of the binder. Since CMC does not strongly interact with the conductive additive, good dispersability was observed compared to conventional binders. In a different study, the effect of CMC on the stability and chemical properties of a natural graphite suspension in an aqueous medium was studied [15]. As for the previous report, a correlation was established between dispersion stability and electrochemical performance. The crucial factor for stabilizing the graphite suspensions was swelling of CMC. Electrochemical experiments revealed that a half cell consisting of Li/organic electrolyte/natural graphite anode and 750 mAh class lithium ion cells exhibited an initial discharge capacity >340 mAh g^{-1} and an improved charge-discharge efficiency. In a different study, graphite in conjunction with Si-NPs anode materials and SBR-CMC as binder was reported. Assembled half cells in EC:DMC (1:1) using 1 M $LiPF_6$ showed that both SBR-CMC and CMC had similar bonding ability as conventional poly(vinylidene fluoride) (PVdF). However, CMC featured a much smaller irreversible charge capacity in the first cycle compared to PVdF binders. The main advantage of the SBR-CMC binder is that an electrode consisting of 1% SBR/1% CMC as binder showed the same cycle stability as an identical electrode containing 10% PVdF binder. In order to evaluate the effect of COOH groups on the dispersability of graphite, four types of cellulose (CMC, HEC, methyl cellulose, ethyl cellulose), were tested as potential binders in graphitic anodes for LIBs [16]. Already a rather small amount of 2% cellulose (derivatives) yielded acceptable anode properties (reversible capacity/300 mAh g^{-1} during the first 10 cycles, irreversible loss B/20%).

Four different binders (Li-CMC, Na-CMC, xanthan gum, PEDOT) for mesocarbon microbeads (MCMBs) anode materials in Li-ion batteries were screened by Courtel [17]. Investigations into thermal stability of the binders revealed by DSC and TGA, showed melting points in a range between 100 and 150 °C, with an onset temperature for decomposition above 220 °C. Li/MCMB half-cell batteries were assembled by incorporating the binders, followed by electrochemical characterization. The cells containing the XG showed the best cycling performance (capacities reaching 350 mAh g^{-1} after 100 cycles at C/12), while the others featured capacities similar to those of the conventional binder PVDF. The optimum thickness of the XG-based MCMB electrodes was determined to be 300–365 μm to give the highest capacities and sustained high C-rates.

A very interesting approach to image the changes during cycling in the three-dimensional (3D) microstructure of a silicon/carbon/CM-cellulose (Si/C/CMC) electrode for Li-ion batteries is investigated by combined focused ion beam/SEM tomography [18]. Particular imaging methods had been applied to reconstruct a volume element of $20 \times 8 \times 11$ $μm^3$ wherein the Si and pore phases were clearly identified before and after 1, 10 and 100 cycles. They showed that the Si particles (size: 0.37 μm) and pores (size 0.40 μm) are homogeneously distributed and fully connected in the native electrode material. During cycling major changes were observed in the morphology of the electrode materials (cracking) and the growth of the SEI was observed along with changes in the particle size and shapes. After 100 cycles, particles with a size 0.14 μm were detected with a non-spherical morphology (4.6 aspect ratio, Fig. 2.4).

Carboxymethyl cellulose (CMC), a low-cost binder [19], is used to make lithium-ion battery composite electrodes containing the high voltage cathode material Li_2MnO_3–$LiMO_2$. This combination of materials results in a homogeneous

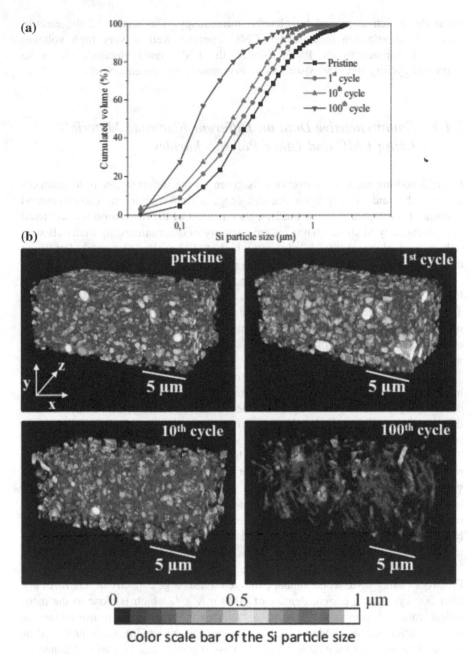

Fig. 2.4 a Distribution curves and **b** 3D views of the Si particle size for the electrodes before cycling and after the 1st, 10th and 100th cycle [18]. Reproduced with permission of the Electrochemical Society

electrode, which is proved by electron microscopy. The results of the electro-chemical investigation indicate that CMC operates well at very high voltages (4.8 V). Compared to the PVDF binder, the CMC-based electrode shows an improved cycling stability as well as a very promising rate capability.

2.1.1 Comprehensive Data on Different Electrode Materials Using CMC and Other Polysaccharides

In the following section, an overview is given on the different electrode materials using CMC and other polysaccharides (e.g., alginate, chitosan, carboxymethyl chitosan, pectin, guar gum) as binders. So far, most of the discussion was centered to high-capacity Si-anodes but there are a variety of other interesting materials with rather good electrochemical behavior which will be also discussed. For more details, the reader shall consult the full papers referenced in this section.

The first example includes lithium titanate ($Li_4Ti_5O_{12}$) as anodic active material, lithium iron phosphate ($LiFePO_4$, LFP) as cathodic active material, CMC as binder and an electrolytic solution based on the non-flammable ionic liquid N-butyl-N-methylpyrrolidinium bis(fluorosulfonyl)imide (PYR14FSI) [20]. At room temper-ature, a specific capacity of 140 mAh g^{-1} was determined which was constant for more than 150 cycles.

Mixed metal oxides have been attracting more and more attention since they have the potential to improve the electrochemical performance of single metal oxides [21]. Potential advantages include structural stability, electronic conduc-tivity, and reversible capacity. Uniform yolk-shelled $ZnCo_2O_4$ microspheres were reported by Li and coworkers [21]. They synthesized this material by pyrolysis of ZnCo-glycolate microsphere precursors which were prepared via a simple refluxing route without any precipitant or surfactant (Fig. 2.5). The formation process of the yolk-shelled microsphere structure was reported to be mainly based on the heterogeneous contraction caused by non-equilibrium heat treatment. The perfor-mances of the as-prepared $ZnCo_2O_4$ electrodes using Na-CMC and PVDF as binders revealed that constant current and rate charge-discharge testing results that the $ZnCo_2O4$ electrodes using CMC show much better behavior.

The binder had superior performance than those using PVDF as the binder. Electrodes using CMC as the binder exhibited a discharge capacity of 331 mAh g^{-1} after 500 cycles at a current density of 1000 mA g^{-1}, which is close to the theo-retical value of graphite (371 mAh g^{-1}). Also manganites of transition and/or post-transition metals, AMn_2O_4 (where A was Co, Ni or Zn), were proposed as anode materials for Li-ion batteries [22]. These materials can be easily obtained by co-precipitation whereas $ZnMn_2O_4$ showed the most promising results during cycling (regarding discharge capacity, cycling, and rate capability) compared to the two other manganites and their corresponding simple oxides. The effect of sintering on particle size was studied in a range from 400 to 1000 °C and revealed a

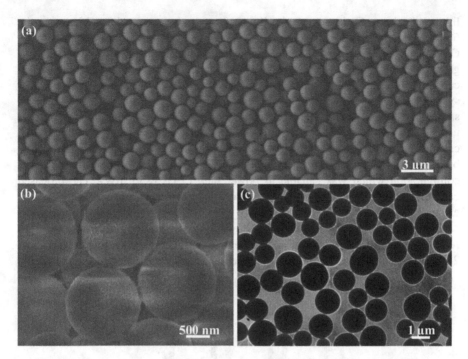

Fig. 2.5 SEM (**a**, **b**) and TEM images (**c**) of the as-synthesized ZnCo-glycolate precursor fabricated by refluxing at 170 °C for 2 h. Reproduced from Ref. [21] with permission of the Royal Society of Chemistry. Copyright © 2013, Royal Society of Chemistry

significant influence on battery characteristics. For example, the optimum particle size for $ZnMn_2O_4$ was in the range between 75 to 150 nm. Both, lithium and sodium salts of CMC, improved the performance of the batteries compared to the conventional binder, PVDF. The best performance was achieved using $ZnMn_2O_4$ powder (sintered at 800 °C, particle size <150 nm) and Li-CMC binder with a capacity of 690 mAh g^{-1} (3450 mAh mL^{-1}) at C/10, and excellent capacity retention (88%). A Mn_3O_4/graphene composite based on graphene platelets and a Mn_3O_4/reduced-graphene-oxide composite were investigated in conjunction with Li-CMC as binder [23]. The Mn_3O_4/graphene-platelet and the Mn_3O_4/reduced-graphene-oxide composites anode system exhibited high gravimetric capacities ($\sim$700 mAh g^{-1}) and excellent cycling stability over more than 100 cycles. $ZnFe_2O_4$ nanoparticles as an anode material for lithium ion batteries have been reported by Zhang et al. [24]. As binder a mixture of SBR and CMC was used (1:1, wt./wt.). A discharge capacity of 873 mAh g^{-1} was reported after 100 cycles at a 0.1C rate, with a rather low capacity fading rate of 0.06% per cycle. It was demonstrated that the SBR/CMC binder improved the adhesion of the electrode film to the current collector, and provided an effective three-dimensional network for electrons transportation. It was shown that the SBR/CMC binder formed a uniform SEI layer around the anode material thereby prohibiting the formation of lithium dendrites. Carboxymethyl cellulose was also investigated as a binder for the 5 V cathode material $LiNi_{0.4}Mn_{1.6}O_4$. Compared with electrodes using PVDF, CMC coated $LiNi_{0.4}Mn_{1.6}O_4$ exhibited a better discharge capacity at all investigated rates. At 0.2 C, the discharge capacity is nearly matching the theoretical capacity (146 mAh g^{-1}) with a low self-discharge (10%). The use of this binder system potentially allows also for the manufacturing of large cells. Yeo et al. [25] studied the scale up using a biobased cathode material, namely lumichrome, in LIBs using SBR-CMC based binders. Large pouch cells were prepared by simple tape casting using lumichrome with an alloxazine structure and aqueous SBR-CMC binders. They assembled a battery module with a 2-in-series, 6-in-parallel (2S6P) configuration which was capable to power blue LEDs (850 mW). The biobased cathode material did not show any alteration in structure during fabrication of the electrode material.

The electrochemical investigations showed that the large pouch cells (Fig. 2.6) featured 2 sets of cathodic and anodic peaks with average potentials of 2.58 and 2.26 V versus Li/Li⁺. The initial discharge capacities were reported to be 142 and

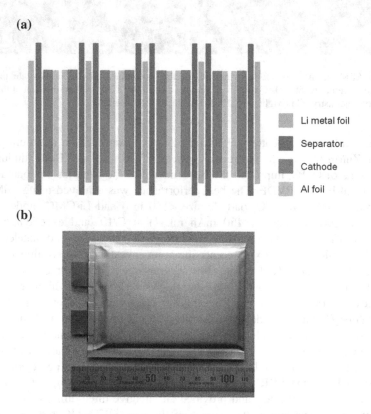

Fig. 2.6 a A schematic diagram of the pouch cell structure and **b** as-prepared pouch cell. Reproduced from Ref. [25] with permission from Elsevier

148 mAh g^{-1} for EC-DMC and tetraethylene glycol di-Me ether (TEGDME) electrolytes, which was similar to those of a coin cell (149 mAh g^{-1}), whereas the EC-DMC-operated pouch cells exhibited higher rate performance and cyclability than those with TEGDME.

In a different report, Li–S batteries with different binders were in the focus [26]. As binders, CMC-SBR, alginate and LA132, a polyacrylic latex, were investigated to study the dispersion mechanism on the cathode materials and the consequent influence on the performance of Li–S batteries. A wide range of experimental and simulation techniques (zeta potential, differential scanning calorimetry analysis and calculations of the rotational barriers of the links of the polymer chains by General Atomic and Molecular Electronic Structure System) revealed that high charge densities and chain flexibility of the binders govern the dispersion of the downsized cathode materials. Although CMC-SBR had a rather good performance, LA132 showed even better dispersion and stabilization of the cathode materials in aqueous environment. Cathodes with better dispersion in the binder led to higher discharge capacities.

Alginate was also proposed in a different study to be used as a binder in the preparation of sulfur cathodes for lithium-sulfur batteries [27]. The EIS tests indicated that the alginate coated sulfur cathode had lower resistance and better kinetic characteristics than those cathodes using PVDF as a binder in an N-methyl-2-pyrrolidone (NMP) solvent. The discharge capacity and the capacity retention rate of alginate sulfur cathode were 508 mAh g^{-1} and 65.4% at the 50th cycle with a current density of 335 mA g^{-1}. The alginate sulfur cathode exhibited much better cycle characteristics than the PVDF coated ones.

A poly (acrylic acid sodium)-grafted-CM-cellulose (NaPAA-g-CMC) copolymer, prepared by free radical graft polymerization from CMC and acrylic acid, was proposed as binder for Si anode materials in LIBs [28]. The grafting with acrylic groups led to an increase of binding contacts with the Si anode materials and the copper current collector while forming a stable SEI on the Si surface. Compared to pure CMC and NaPPA, the NaPAA-g-CMC based Si anode exhibited much better cycle stability and higher coulombic efficiency. In a different report, a PAA-CMC binder for Si anodes yielded 2000 mAh g^{-1} after 100 cycles at 30 °C while maintaining a high capacity and high current density [29].

$LiNi_{0.33}Mn_{0.33}Co_{0.33}O_2$ (NMC) as a cathode material for lithium ion batteries was reported by Xu (Fig. 2.7) [30]. The material was synthesized using a sol-gel approach and the X-ray diffraction Rietveld refinement results indicated that a single-phase NMC with hexagonal layered structure was obtained, with uniform particle sizes in the range of 100–200 nm as proven by SEM. The performance of the NMC electrodes with CMC, PVDF, and alginate from brown algae as binders was compared. NMC electrode using CMC as binder featured the highest rate capability, followed by those using alginate and PVDF binders, in constant current charge–discharge tests.

Electrochemical impedance spectroscopy showed that the electrode using CMC as the binder had lower charge transfer resistance and lower apparent activation energy than the electrodes using alginate and PVDF as the binders. The apparent

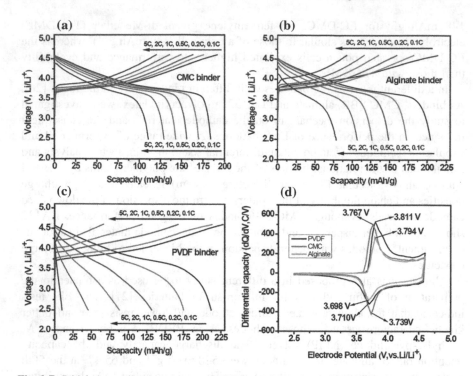

Fig. 2.7 Initial charge discharge curves of NMC at various current densities in the voltage range of 2.5–4.6 V using different binders: **a** CMC, **b** alginate, and **c** PVDF; **d** dQ/dV curves for NMC with CMC, PVDF, and alginate binders. Reproduced from Ref. [25] with permission from Elsevier

activation energies of NMC electrodes using CMC, alginate, and PVDF as binders were calculated to be 27.4, 33.7, and 36 kJ mol^{-1}, respectively.

CMC was also used in the manufacturing of anatase TiO_2 electrodes for LIBs [31]. The low temperature performances at different charge/discharge rates of TiO_2/CMC and TiO_2/PVDF electrodes were compared and yielded much better performance for the CMC binder compared to commercial PVDF ones. Similarly, the same materials was proposed for fuel cell applications with a $LiFePO_4$ cathode [32]. The assembled fuel cell showed high stability in terms of cycling stability and charge/discharge profiles.

Zhang and coworkers [33] compared different binders (CMC, PVDF and PAA) for $LiMn_2O_4$ cathodes and their influence on the adhesion strengths, swelling properties, morphologies and electrochemical behavior were studied. $LiMn_2O_4$ cathodes with PAA/NMP system displayed the best cycle performances at both 25 C and 55 C among these four cathodes, whereas the better capacity retention for $LiMn_2O_4$ cathode with PAA/NMP system was related to strong binding ability, appropriate swelling property and homogeneous distribution of particles inner the electrode. Two different types of binders, CMC and CMC-formiate (CMC-f) were investigated in respect to the cyclability of MgH_2 based electrode systems [34]. These electrodes exhibit a large reversible capacity of 1800–1900 mAh g^{-1} at an average voltage of 0.5 V versus $Li^+/Li°$. This is a suitable range for usage as anode materials in LIBs. Moreover, addition of carbon to the binders improved capacity retention (240 and 542 mAh g^{-1}, respectively).

Carboxymethyl chitosan (CM-Ch) was reported as binder for $LiFePO_4$ cathode in Li-ion batteries by Sun and coworkers [35] and its electrochemical performance was compared to CMC and PVDF. The $LiFePO_4$ cathodes with CM-Ch binders exhibited a better rate capability than those with CMC and PVDF, retaining 65% capacity of C/5 at 5 C rate as compared with 55.9 and 39.4% for CMC and PVDF, respectively. In addition, the cycling performance at 60 °C with CM-Ch showed good behavior, retaining 91.8%/62.1% capacity after 80 cycles at 1 C/10 C, respectively. In a different study, silicon nanoparticles were investigated in combination with CM-Ch [36]. Fourier transformation infrared spectroscopy (FTIR) and X-ray photoelectron spectroscopy (XPS) measurements revealed that strong hydrogen bonding was present between the hydroxylated Si surface and the polar groups (–OH, –COOH and $–NH_2$) of CM-Ch. The Si/C-Cm-Ch anode (Si:carbon black:CM-Ch = 62:30:8 wt./wt./wt.) exhibited a high first discharge capacity (4270 mAh g^{-1}) with a first coulombic efficiency of 89%, and maintained a capacity of 950 mAh g^{-1} at the current density of 500 mA g^{-1} over 50 cycles.

A reactive binder was proposed by He et al. [37]. They synthesized cyanoethylated carboxymethyl chitosan (CN-CM-Ch) by a straightforward cyanoethylation reaction of CM-Ch with acrylonitrile in aqueous NaOH. The adhesion strength of CN-CM-Ch on $LiFePO_4$ improved to 0.047 N/cm (compare to 0.013 N/cm for CM-Ch) after introducing the cyanoethyl group. The electrochemical performance of $LiFePO_4$ electrode with CN-CM-Ch exhibited improved cycling stability and rate capability, retaining 56.3% capacity of C/5 at 5C rate as compared with 48.4 and 32.8% for CMC and PVDF, respectively. Additional electrochemical characterization (CV, impedance spectroscopy) revealed that CN-CM-Ch on $LiFePO_4$ electrodes led to a more favorable electrochemical kinetics than those with CMC and PVDF.

CMC-Li was used as a binder for 9,10-anthracenedione (AQ) electrodes [38]. The AQ electrodes were investigated by galvanostatic discharge/charge, cyclic voltammetry and electrochemical impedance spectroscopy techniques. At room temperature, the CMC-Li electrode showed better electrochemical performance compared to PVDF electrode, exhibiting a specific capacity of up to 214 mAh g^{-1} at the initial discharge, and its specific capacity was maintained at 62 mAh g^{-1} after 50 cycles. In addition, better stability was achieved during the charge and discharge processes. Furthermore, the electrochemical performance of the CMC-Li with a $DS_{COOH} = 1.0$, was superior to those having lower DS_{COOH} (0.62).

A hard-carbon negative electrode with a CMC binder demonstrated superior reversibility and cyclability in non-aqueous sodium-ion batteries [39]. As electrolyte, $NaPF_6$ in propylene carbonate was employed at room temperature and compared to PVdF. Furthermore, effects of monofluoroethylene carbonate (FEC) additives remarkably depend on the combination with binders, CMC and PVdF. Surface analyses revealed considerable differences in surface and passivation

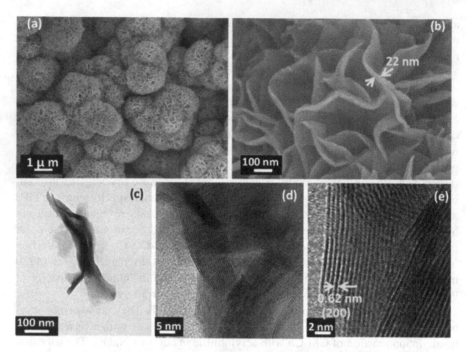

Fig. 2.8 FEG-SEM (**a, b**) and FEG-TEM (**c–e**) images of MoS$_2$ nanowall morphology. Reproduced from Ref. [40] with permission of the American Chemical Society

chemistry which depends on the binders and FEC additive used for the hard-carbon negative electrodes.

CMC was investigated as binder in electrochemically stable molybdenum disulfide (MoS$_2$) electrodes featuring a two-dimensional nanowall structure (Fig. 2.8) [40].

The MoS$_2$ was prepared by a simple two-step synthesis method followed by thermal annealing at 700 °C in a reducing atmosphere. CMC featured a better electrochemical performance and stability of the MoS$_2$ nanowalls compared to PVDF. The electrodes exhibited a high specific discharge capacity of 880 mAh g^{-1} at 100 mA g^{-1} without any capacity fading for more than 50 cycles. Further, an outstanding rate capability with a reversible capacity have been reported with values as high as 737 and 676 mAh g^{-1} at rates of 500 and 1000 mA g^{-1} at 20 °C, respectively. The authors attributed the excellent electrochemical stability and high specific capacity of the nanostructured materials to the two-dimensional nanowall morphology of MoS$_2$ in combination with the CMC binder. In a similar manner, ultralong α-MoO$_3$ nanobelts (average length: 200–300 μm) with uniform width of around 0.6–1.5 μm have been used in conjunction with CMC binders [41]. The nanobelts were synthesized by a hydrothermal method using a molybdenum organic salt precursor. As for the previous example, the CMC led to much better electrochemical performance than those electrodes containing PVDF binders.

Remarkably, the electrodes containing CMC exhibited high specific capacity of over 730 mAh g^{-1} for over 200 cycles at a 0.2 C rate. At rates of 1–2 C, high capacities of around 430–650 mAh g^{-1} were reported. Besides CMC, also alginates were tested as binders in these studies and they revealed stable capacity retention of around 800 mAh g^{-1} for over 150 cycles at 0.2 C as well.

Tin nanoparticle/polypyrrole (nano-Sn/PPy) composites were used in combination with CMC and PVDF as binders [42]. The prepared composites showed a much higher surface area than the pure nano-Sn reference sample. The authors argue that the porous higher surface area of PPy and the much smaller size of Sn in the nano-Sn/PPy composite than in the pure tin nanoparticle sample is responsible for this behavior. The electrochemical investigations of all the materials revealed that both, the capacity retention and the rate capability, are in the same order of nano-Sn/PPy-CMC > nano-Sn/PPy-PVDF > nano-Sn-CMC > nano-Sn-PVDF.

Carboxymethyl chitosan (CMCh) and chitosan lactate (ChLac), were employed as binders for micro- and nanostructured SnS$_2$ electrodes and compared to CMC and non-aqueous PVDF [43]. All the electrodes prepared by using the water-soluble binders (CMCh, ChLac and CMC) featured higher initial coulombic efficiency, larger reversible capacity, and better rate capabilities than those with PVDF. Although the SnS$_2$ electrodes with CMCh as binders had a better rate capability at a high rate of 5C, they showed a slightly worse cycling stability than CMC.

CMC, guar gum (GG), and pectins have been investigated as binders for the manufacturing of lithium titanate (Li$_4$Ti$_5$O$_{12}$, LTO) electrodes [44]. Phosphoric acid was added during electrode slurry preparation to prevent dissolution of the aluminum current collector. Without phosphoric acid, hydrogen evolves during coating of the slurry onto the aluminum current collector, concomitant with the formation of cavities in the coated electrode and poor cohesion on the current collector itself. As a consequence, the addition of phosphoric acid to the slurries significantly improved the electrochemical performance of the electrodes. At a 5 C rate, CMC/PA-based electrodes provided 144 mAh g^{-1}, while PA-free electrodes performed worse (124 mAh g^{-1}). Although GG and pectin showed a slightly worse adhesion to the current collector, the electrodes featured comparable electrochemical performance than those based on CMC. Full lithium-ion cells, utilizing CMC/PA-made LiNMC cathodes and LTO anodes provided a stable discharge capacity of $\sim$120 mAh g^{-1} (NMC) with high coulombic efficiencies. Besides GG, tara gum (TG), a galactomannan derived from plant seeds, was used as the binders for LTO anodes in LIBs [45]. Although their adhesion on the electrode materials was not as strong as for CMC, both galactomannan gums facilitated the transport of lithium ions in LTO electrodes compared to CMC binders, which was probably due to their branched structure. The branched structures were proposed to accommodate large amounts of electrolyte. In terms of electrochemical performance, the GG-coated LTO electrode featured a high reversible capacity of 160 mAh g^{-1} after 100 cycles at 1 C current rate (compare CMC: 150.1 mAh g^{-1}). At higher current rates, the difference in capacity between the GG and CMC electrodes increased and reached 25 mAh g^{-1} at 10 C current rate.

Synthetic graphite (SG)-based electrodes of LIBs were used in combination with different binders, namely PVDF, CMC, alginate, gum arabic (GA), XG, GG, agar-agar (AA) and caragenaan (CG) [46]. All chosen binders were electrochemically and thermally stable under the employed experimental conditions. For SG/hydrocolloid electrodes, binder concentrations of 5 wt% were chosen. Good to excellent electrochemical performances for electrodes with alginate, CMC, XG and GG in galvanostatic cycling experiments were obtained at constant (C/10, with $C = 372$ mA g^{-1}) and variable (from C/10 to 2C) current rates, which were superior to those of SG/PVDF electrodes with higher binder content (8 wt%). In contrast, SG/GA, SG/CAR and SG/AA electrodes featured poorer electrochemical performances (Fig. 2.9), which was probably related to the low adhesion capacity of the binder (GA and CAR), or the formation of films covering the SG particles (CAR and AA).

XG was proposed by Chen et al. [47] as binder for a nanocomposite consisting of silicon nanoparticles dispersed on conducting graphene (Si/graphene) in LIBs. The nanocomposite was synthesized by high-energy ball milling followed by thermal treatment. The Si/graphene composite anode provided an enhanced reversible capacity, excellent cyclic performance and rate capability, compared to the neat Si-anodes. The nanocomposite anode with XG binders behaved much better in terms of cycling and rate performances compared to electrodes prepared

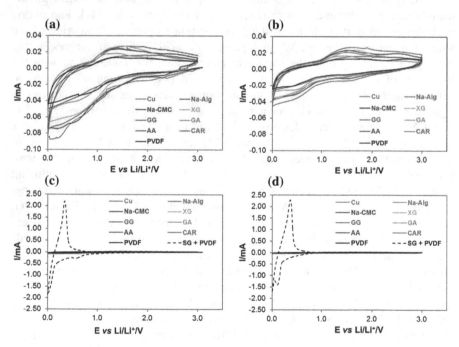

Fig. 2.9 CV voltammograms of cycle 1, (**a** and **c**), and cycle 5 (**b** and **d**), of the binder electrodes and bare Cu (**c** and **d**) also include the SG/PVDF electrode for comparison purposes. Reproduced with Permission from Ref. [46] with permission of Elsevier

using CMC as binder. The enhancement was related to large binder stiffness and strong adhesion of the binder to Si-based particles contributing to maintain the integrity of the electrode and accommodated the volume change of Si during charge/discharge. Jeong et al. [48] provided an analogy of millipede adhesion to battery binders on the basis of XG and other polysaccharides. They reported that the double helical superstructure with side chains and ion-dipole interactions was responsible for the excellent electrochemical performance compared to other polysaccharide binders. In a different study, GA was applied as a binder for Si-anodes in LIBs [49]. The resultant Si anodes had an outstanding capacity of ca. 2000 mAh g^{-1} at a 1 C rate and 1000 mAh g^{-1} at 2 C rate, respectively, throughout 500 cycles. Excellent long-term stability was demonstrated since the materials still provided 1000 mAh g^{-1} specific capacity at 1 C rate after 1000 cycles. GA was also proposed as a binder for the sulfur cathode in Li–S batteries [50]. One of the major advantages of GA in Li–S batteries was their ability to confine sulfur species via its functional groups while providing good mechanical properties. The cycling performance was good and a capacity of 841 mAh g^{-1} at low current rate of C/5 throughout 500 cycles was reported. A comparison between PVDF and GG binders for composite anode electrodes was reported by Kuruba and coworkers [51]. The composite consisted of 82 wt% Si/C lithium ion active material, 8 wt% polymeric binder and 10 wt% Super P conductive carbon black and was synthesized using high energy mechanical milling. The resulting materials exhibited reversible specific capacities of 780 and 600 mAh g^{-1} at charge/discharge rates of ~ 50 and ~ 200 mA g^{-1}, respectively.

By mixing Si nanopowder with alginate high-capacity silicon (Si) nanopowder–based lithium (Li)–ion batteries with improved performance characteristics were reported [52]. The results showed that the obtained materials performed much better than PVDF and CMC in nearly all electrochemical parameters such as charge discharge capacity, retention capacity, Coulomb efficiency and electrode fading to mention just the important ones. This approach was slightly altered in another report where catechol groups were grafted on various types of polysaccharides.

A similar picture was obtained by Feng et al. [53] who investigated the effect of different binders on CdO nanoparticles and CdO/Carboxylated multiwalled carbon nanotubes (CNTs) nanocomposite electrodes for LIBs. The nanosized CdO was capable to deliver an initial capacity as high as 805 mAh g^{-1} with, however, poor retention capacity. By incorporation of CNTs and an alginate binder, the CdO delivered a reversible capacity of 810 mAh g^{-1} for more than 100 cycles and a capacity of 720 mAh g^{-1} at a rate of 1500 mA g^{-1}.

Zhang and coworkers [54] employed a self-assembly strategy to construct a three-dimensional (3D) polymeric network containing alginates for high-performance silicon submicro-particle (SiSMP) anodes. This was accomplished by crosslinking alginate chains by addition of calcium cations using the Eggbox model approach (Fig. 2.10) [55]. As a consequence of crosslinking, the highly cross-linked alginate networks exhibited superior mechanical properties and further showed strong interaction with SiSMPs. Moreover, this network was capable to tolerate the volume change of SiSMPs thereby effectively maintaining the mechanical and electrical

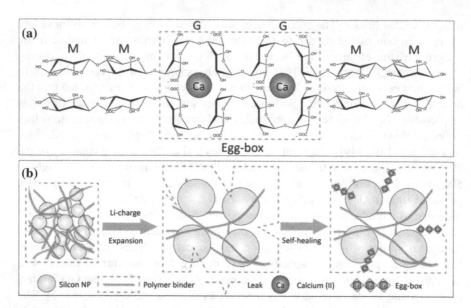

Fig. 2.10 a Molecular structure of alginate and Ca-mediated "egg-box" like cross-links in Ca-alginate. **b** Expansion and self-healing mechanism of Ca-alginate-containing silicon anodes during charge–discharge cycles. Reproduced from Ref. [55] with permission of the Royal Society of Chemistry

integrity of the electrode and significantly improving the electrochemical performance. Similar as for the CMC, a self-healing mechanism was proposed [55]. As a result, SiSMPs with a 3D binder network exhibit high reversible capacity, superior rate capability and much prolonged cycle life. Similar findings were obtained by Liu et al. they also used Ca^{2+} as crosslinker for alginates to obtain stable alginate binders for Si/C electrodes. Materials with remarkably improved electrochemical properties have been reported [56].

Alginates were also employed as binders in nanocomposites composed of $CoFe_2O_4$ nanoclusters with different concentrations of reduced graphene oxide (rGO) [57]. The electrodes containing $CoFe_2O_4$ + 20% rGO composite with alginate as binder revealed high capacity of 1040 mAh g^{-1} at 0.1 C (91 mA g^{-1}) rate with excellent rate capability, which was much higher than those of PVDF binders. In a different study, a nanoflaky MnO_2-graphene sheet (GS) hybrid material with an alginate binder was used as a cathode for LIBs [58]. The MnO_2 growing on the GS led to high capacity of 230 mAh g^{-1} at a current density of 200 mA g^{-1}, even after more than 150 cycles. Alginates, were also used as binder for spinel type $LiMn_2O_4$ electrodes [59]. Since alginates strongly interact with divalent cations such as Mn^{2+}, the potential of LMO could be fully exploited. The authors provided some proofs of concept of using this material in LiBOB half and full cells.

Alginate binders have been reported also in conjunction with hollow nanostructural features α-Fe_2O_3 nanotubes for usage in LIBs [60]. In this regard, α-Fe_2O_3

was hydrothermally synthesized to give hollow nanostructured α-Fe_2O_3 nanotubes. This electrode material exhibited high capacities (800 mAh g^{-1} at 503 mA g^{-1} for 50 cycles) when an alginate binder was employed. Even at higher current rates (e.g. 1007 mA g^{-1}), high capacities have been reported (732 and 600 mAh g^{-1}) after 50 and 100 cycles, respectively. The same electrode setup also featured good rate capability and provided a capacity of 400 mAh g^{-1} even at a very high current density of 10 A g^{-1}. Similar to CMC, weak hydrogen bonding between the surface hydroxyl groups on the metal oxide (Fe_2O_3) and the carboxylic functional groups on the alginate binder was suggested for the enhanced battery performance at very high current rates.

A crosslinked chitosan derivative was proposed as binder for antimony anode materials in sodium batteries [61]. The crosslinking of the chitosan was performed using glutaraldehyde to give a three dimensional network which was capable to accommodate volume changes upon sodium insertion/removal. Similar as CMC, chitosan also provided a stable SEI. The electrochemical performance was improved by crosslinking the chitosan, leading to a capacity of 555.4 mAh g^{-1} at 1 C after 100 cycles with a capacity retention of 96.5% compared to the 1st cycle. For comparison, the charge capacity of the electrode without the crosslinking step for chitosan was lower with 463.4 mAh g^{-1} after 100 cycles with a capacity retention of 81.3%. The same crosslinking strategy can also be applied for LIBs and Si anode materials [62]. As for the previous example, a 3D network was built up to limit the movement of Si particles through the cross-linking between the amino groups of CS and the dialdehyde of glutaraldehyde. The crosslinked anode material exhibited an initial discharge capacity of 2782 mAh g^{-1} with a high initial Coulombic efficiency of 89% and maintained a capacity of 1969 mAh g^{-1} at the current density of 500 mA g^{-1} over 100 cycles.

Chen and coworkers [63] showed that chitosan, either as additive in separators or cathode materials, was capable to trap polysulfides in Li–S batteries. Cathodes with chitosan featured enhanced initial discharge capacities (950–1145 mAh g^{-1} at C/10) while reversible specific capacity after 100 cycles increased from 508 to 680 mAh g^{-1} and 473 to 646 mAh g^{-1} at rates of C/2 and 1 C, respectively. Batteries with separators having a carbon/chitosan layer exhibited high discharge capacity (830 mAh g^{-1} at C/2 after 100 cycles and 675 mAh g^{-1} at 1 C) after 200 cycles. The capacity fading was determined to be 0.11% per cycle.

Spherical graphite anodes were also equipped with chitosan as a binder for LIBs [64]. With similar specific capacity, the first Columbic efficiency of the chitosan-based anode was 95.4% compared to 89.3% of a PVDF-based anode. After 200 charge–discharge cycles at 0.5C, the capacity retention of the chitosan-based electrode showed to be significantly higher than that of the PVDF-based electrode.

Besides highly polymeric chitosan, chitosan oligosaccharides (COS) have been proposed as a new, environmentally benign and water-based organic compounds, to be used as electrode binder for electrodes in LIBs [65]. $Li_2ZnTi_3O_8$ electrodes coated with COS exhibited a significant improvement of the electrochemical performance in terms of the first Columbic efficiency, cycling behavior, rate capability and life cycle. The initial discharge capacity was 215.6 mAh g^{-1} at 0.1 A g^{-1}, with a Columbic efficiency of 93.6%. After 1000 cycles, capacity was determined to be 66.1 mAh g^{-1} and the retention was reported 33.6%.

Jeong et al. [66] introduced polymerized β-cyclodextrin (β-CDp) as binder for Si nanoparticle anodes. The inherent highly branched structure of β-CDp provided a variety of hydrogen bonding modes and interaction sites with Si particles and therefore offered robust contacts between the individual components. As for CMC, self-healing took place, meaning that binder-Si nanoparticles interactions recover during cycling, thereby maintaining good electrode performance. As one of the essential components in electrodes, the binder affects the performance of a rechargeable battery. By partial oxidation of β-cyclodextrin (β-CD) using hydrogen peroxide, a new binder, C-β-CD, for sulfur composite cathodes was obtained [67]. A major advantage was its water solubility at room temperature which is ca. 100 times higher than that of β-CD. C-β-CD features all the required properties of an aqueous binder: strong bonding strength, high solubility in water, moderate viscosity, and wide electrochemical windows. C-β-CD coated sulfur composite cathodes were reported to have high reversible capacity of 694.2 mAh g(composite)$^{-1}$ and 1542.7 mAh g(sulfur)$^{-1}$, with a sulfur utilization of ca. 92%. The discharge capacity remained at 1456 mAh g(sulfur)$^{-1}$ after 50 cycles, which is much higher than that of the cathode with unmodified β-CD as binder.

In a different approach, supramolecular cross-linking via dynamic host–guest interactions between hyperbranched β-cyclodextrin polymer and a dendritic gallic acid cross-linker incorporating six adamantane units was used for high-capacity silicon anodes (Fig. 2.11) [68]. The authors proved by calorimetry in the solution phase that the given host–guest complexation is a highly spontaneous and enthalpically driven process, which was substantiated by gelation experiments in both aqueous and organic media. The dynamic cross-linking process allows for more efficient silicon–binder interactions, provides structural stability of electrode films, and creates the formation of a defined electrode–electrolyte interface. The impact of the dynamic cross-linking was maximized at an optimal stoichiometry between the two components. Importantly, the present investigation proved that the molecular-level tuning of the host–guest interactions could be directly translated to the cycling performance of silicon anodes.

Molecular structures of polysaccharide binders determining mechanical properties were correlated to electrochemical performances of silicon anodes for lithium-ion batteries. Glycosidic linkages (α and β) and side chains (–COOH and –OH) were selected and proven as the major factors of the molecular structures. Three different polysaccharides (CMC, pectin and amylose) were investigated by Yoon et al. [69] in respect to their performance as binders for silicon anodes. Pectin was remarkably superior to CMC and amylose in cyclability and rate capability of battery cells based on silicon anodes. The pectin binder allowed for volume expansion of silicon electrodes while maintaining high porosity during lithiation and the physical integrity of pectin-based electrodes was not altered during repeated lithiation/delithiation cycles. Murase et al. [70] compared amylose, amylopectin and glycogen in terms of their binder efficiency for Si-anode materials and aimed at

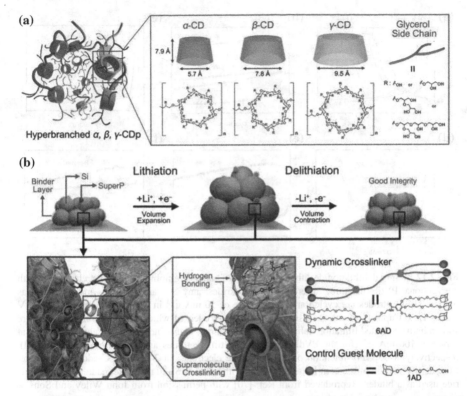

Fig. 2.11 a Graphical representation and chemical structures of hyperbranched, β, and γ-CDp. **b** Proposed working mechanism of dynamic cross-linking of β-CDp and AD in an electrode matrix along with graphical representations and chemical structures of guest molecules incorporating adamantane moiety. Reproduced from Ref. [68] with permission of the American Chemical Society

correlating branching to binder performance. As shown in Fig. 2.12, amylose, a linear polysaccharide, performed worse than the branched amylopectins and glycogen, which the authors related to the better adhesion to the electrode materials of the latter.

The potential applicability of agarose as an electrode binder and also as a carbon source for high-performance rechargeable lithium-ion batteries was reported by Hwang et al. [71]. The agarose binder facilitated adhesion of silicon (Si) active materials to copper foil current collectors. As a consequence, a significant improvement in the electrochemical performance of the resulting Si anode (specific capacity = 2000 mAh g^{-1} and capacity retention after 200 cycles = 71%) was achieved. In addition, agarose was exploited as a cathode binder for LMO featuring excellent cell performance (initial coulombic efficiency of 96.2% and capacity retention after 400 cycles of 99%). Further, by selective carbonization of Si-dispersed agarose, Si/C (hard carbon) composite active materials were obtained. Eventually, the Si/C composite anode and the LMO cathode mentioned above were

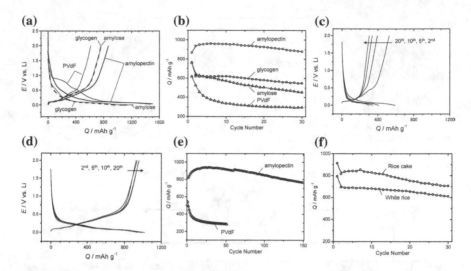

Fig. 2.12 a Initial lithiation/de-lithiation curves of nano-Si/graphite composite electrodes with conventional PVdF and three different polysaccharides, amylose, amylopectin, and glycogen, as binder. The electrodes are cycled at a current rate of 50 mA g^{-1} in a voltage range of 0.0–2.0 V versus Li. **b** Capacity retention of the composite electrodes, of which the first voltage curves are shown in (**a**). Selected lithiation/delithiation voltage curves at the 2nd, 5th, 10th, and 20th cycles at a rate of 100 mA g^{-1} for the PVdF and amylopectin electrodes are also shown in (**c**) and (**d**), respectively. **e** Capacity retention of the composite electrodes in 2 vol% FEC-added electrolyte. **f** Capacity retention of the composite electrodes with white rice and rice cake made from glutinous rice used as a binder. Reproduced from Ref. [70] with permission from John Wiley and Sons

assembled to produce a full cell featuring the use of agarose as an alternative green material. The full cell showed a stable cycling performance (capacity retention after 50 cycles of >87%).

2.2 Polysaccharides as Separators

Separators are crucial parts of any type of battery since they provide a physical barrier between the anode and cathode to avoid short circuits [72]. Additionally, they act as storage medium to provide electrolyte for the transport of ions during charge and discharge. As a consequence, separators are not directly involved in cell reactions but their morphology, structure and physical parameters strongly influence battery performance parameters. In principle, separators must be inert towards the electrolyte and the electrode materials, even under strong oxidizing and reducing conditions. They should not act as catalysts, thereby producing impurities which may reduce battery performance. For some applications, even stability towards elevated temperatures at corrosive conditions must be provided by separator materials. In order to have low internal resistance and high ionic conductivity, the electrolyte must easily absorb into the separator materials.

In fact there are five major classes of separators, namely microporous membranes, modified microporous membranes (e.g., by surface grafting), non-woven mats, composite membranes and electrolyte membranes. The electrolyte membranes can be divided into solid polymer electrolytes (SPE), and gel polymer electrolytes (GPE). It is obvious that depending on the envisaged applications, certain material designs will be preferred over others and in the following approaches polysaccharides in the different types of separators will be discussed in detail. For a more comprehensive overview on separators in general the reader is referred to recent review articles [72, 73].

2.2.1 Microporous Membranes

Cellulose based separators have been used since ages in the design of batteries. This includes classic Leclanche cells (Zn–C), lead acid batteries, and alkaline batteries. In the past years, the emergence of new cellulose based nanomaterials extended the scope of microporous membranes based on cellulose in battery separator applications. A rather old example is described by researchers from Asahi Chemical Industries [74] where a regenerated microporous cellulose film (pore sizes: 10–200 nm, thickness 39–85 μm) was used as support for fine cellulosic fibers (500 nm to 5.0 μm). This composite membrane was soaked in ethylene carbonate and other aprotic solvents, and exhibited acceptable physical strength while pinholes, a major issue for cellulose based separators, could not be detected. Further, the complex impedance was found to be at least equal or even lower compared to conventional polyolefin separators for LIBs. Lab-scale cells were manufactured, composed of $LiCoO_2$/petroleum coke electrodes, and as electrolyte a 1.0 M solution of $LiBF_4$/propylene carbonate/ethylene carbonate/γ-butyrolactone was used (25:25:50 v/v/v). The cellulosic separators featured a remarkable initial discharge capacity and retention over 41 charge/discharge cycles and could compete with polyolefin based membranes.

Later special types of paper have been reported to be used as separators in LIBs such as rice paper (RP) [75]. It was composed of interpenetrated cellulose fibers (diameter: 5–40 μm) which built up a highly porous scaffold. Interestingly, the rice paper was electrochemically stable at potentials below 4.5 V versus Li+/Li. A variety of electrode materials such as graphite, $LiFePO_4$, $LiCoO_2$ and $LiMn_2O_4$ have been used to evaluate the compatibility of RP in LIBs and to compare them with commercial polypropylene/polyethylene/polypropylene separator membranes. The RP separators having a similar thickness exhibited a lower resistance than the commercial separator. The authors argued that the flexibility, high porosity, low cost and excellent electrochemical performance of the RP membrane may potentially replace commercial separators in lithium-ion batteries for low power applications.

In two different approaches [76], cellulose nanopapers from cellulose nanofibrils (CNF) have been manufactured. CNFs, which can easily isolated from many types of cellulosic feedstock, are characterized by the presence of individual fibrils in the nanometer range which are up to several micrometers long while being highly crystalline. Therefore they contribute to excellent mechanical/thermal properties and support the formation of nanoporous structures evolution. CNF membranes are prone to soak a variety of electrolytes whereas the porous structure can be fine-tuned by variation of the solvent mixtures, here isopropanol-water. It was shown that the separator characteristics, and the electrochemical performance of cells assembled with the CNF separators worked best at IPA:water ratios of 95:5 (v/v%). This approach was extended by addition of colloidal SiO_2 nanoparticles to control the formation of the porous structure which can be challenging for pure cellulose nanopaper separators from densely-packed CNFs. The main feature of this new material was that the incorporated SiO_2 nanoparticles acted as disassembling agent to separate the CNF. Therefore, a rather loose packing of the CNF was realized, thereby creating a more porous structure. The unusual pore structure of this material can be fine-tuned by variation of the SiO_2 contents in the CNF suspensions. Notably, the separator manufactured with 5 wt% SiO_2 content exhibited the highest ionic conductivity (Fig. 2.13).

The authors argued that this was a result of a well-balanced combination of nanoporous structure and separator thickness, thus contributing to excellent cell performance. In a similar way, the pore structure of cellulose diacetate (CDA)-SiO_2 composite films have been used to improve the electrolyte wettability and the thermal stability of separators [77]. For CDA, a SiO_2 content of 9.4% was determined to achieve the best results.

The motivation of Leijonmarck and coworkers [78] was to realize mechanically flexible and strong batteries for different power applications such as active radio-frequency identification tags and bendable reading devices. They proposed a method for making flexible and strong battery cells, which were integrated into a single flexible paper structure. CNF was used whereby it acted as both, electrode binder material and separator material. The battery papers were manufactured by a paper-making type process consisting of sequential filtration of water dispersions which contained the battery components. Papers with a thickness of 250 μm were obtained, which featured strength at break of up to 5.6 MPa when soaked in battery electrolyte. The cycling performances showed a reversible capacity of 146 mAh g^{-1} $LiFePO_4$ at C/10 and 101 mAh/g $LiFePO_4$ at 1C which corresponds to an energy density of 188 mWh g^{-1} of the full paper battery at C/10.

Cellulose nanocrystals (CNCs) obtained from *Cladophora* have been used to manufacture separators as well due their very high crystallinity, good thermal stability while being mechanically robust [79]. Separators based on CNCs were made by a paper-making like process involving vacuum filtration, resulting in sheets with a thickness of 35 μm, an average pore size of about 20 nm, and a Young's modulus of ca. 5.9 GPa. After soaking with 1 M $LiPF_6$ EC: DEC (1/1, vol./vol.) electrolyte, an ionic conductivity of 0.4 mS cm^{-1} were determined. These separators were thermally stable up to 150° and electrochemically inert in a

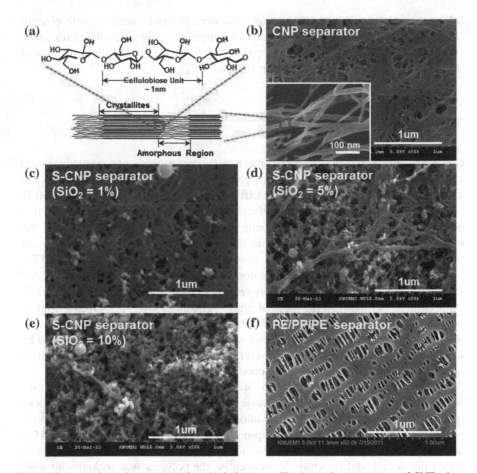

Fig. 2.13 a A schematic representation depicting crystalline/amorphous structure of CNFs that comprise close-packed polysaccharide chains with β-$(1 \rightarrow 4)$-D-glucopyranose repeat units. FE-SEM photographs of: **b** CNP separator, where an inset indicates a high-magnification image; **c** S-CNP separator (SiO$_2$ = 1 wt%); **d** S-CNP separator (SiO$_2$ = 5 wt%); **e** S-CNP separator (SiO$_2$ = 10 wt%); **f** PP/PE/PP separator. Reproduced from Ref. [76] with permission of Elsevier

potential range between 0 and 5 V versus Li+/Li. The cycling stability of a LiFePO$_4$/Li cell with a CNC based separator exhibited good cycling stability having 99.5% discharge capacity retention after 50 cycles at a rate of 0.2 C.

In another study [80], bacterial cellulose nanofibers (BC) have been shown to outperform Celgard 2500 as separator material for Li–S batteries. Separators based on bacterial nanocellulose have a high concentration of nanopores and nanofibers in a rechargeable battery cell allows for safe plating of a dendrite-free metallic-lithium anode. The commercially available bacterial nanocellulose is capable to retain organic-liquid electrolytes longer than Celgard 2500. Further, it featured higher thermal stability, and can be wet by metallic lithium. The authors demonstrated that cells prepared from the membranes as separators showed excellent, safe electro-chemical performance of Li-anode cells and organic liquid electrolytes for up to 1000 cycles. It was also shown that separators from BC exhibit rather high thermal stability and high mechanical strength [81].

2.2.2 Composite Membranes

CNCs have been used to increase the mechanical performance of a commonly used copolymer, PVDF-co-hexafluoropropylene (PVDF-HFP), which has received interest over the years in the area of LIB separator technology [82]. PVDF-HFP/CNC nanocomposite films were manufactured and characterized. The incorporation of CNCs significantly improved Young's modulus and tensile strength of the membranes. The authors proposed that the resulting enhancement of mechanical properties of PVDF-HFP copolymers upon addition of CNCs makes PVDF-HFP a potential candidate for polymer separators in LIBs. In a similar approach, CNC suspensions in DMF were used to improve the mechanical properties of PVDF [83]. Film casting and non-solvent induced phase separation were employed to obtain porous and dense nanocomposite membranes. The CNCs led to a mechanical reinforcement and a lower strain at break while thermal properties also improved by CNC incorporation. As for the PVDF-HFP copolymers, the authors proposed these reinforced membranes to be suitable candidates for separators in LIBs.

A different approach to improve PVDF based separators performance is to use a slurry composed of Al_2O_3, polyvinylidene difluoride-hexafluoropropylene (PVdF-HFP) and carboxymethyl cellulose (CMC) which was deposited on a polyolefin (PE) substrate to form a prototype Al-PHC/PE separator [84]. After assembly and hot-pressing, the PVdF-HFP copolymer was granulated and trans-ferred into a colloidal structure. Subsequently, the polymer was soaked with elec-trolyte, capable to crosslink the Al_2O_3 nanoparticles, thereby increasing battery hardness (Fig. 2.14). The ionic conductivity (9.3×10^{-4} S cm^{-2}) of the Al-PHC/PE-2 separator (12 μm, with a Al_2O_3/PVdF-HFP weight ratio of 7/3) can compete with a PE separator (9 μm), and showed thermal stability up to 110 °C. The capacity retention rose from 84 to 88% by using the Al-PHC/PE-2 instead of the PE separator.

Another type of composite membranes, cellulose/polysulfonamide, have been also reported [85]. These membranes were synthesized by combining microfibril-lated cellulose and polysulfonamide in a papermaking-like process. These mem-branes were investigated towards their applicability as separators in LIBs by characterizing the electrolyte wettability, heat tolerance, and electrochemical parameters. Batteries with lithium cobalt oxide/graphite using the separator featured better capacity retention ratios of 85% after 100 cycles and superior rate capability compared with commercially based polypropylene separators. Other chemistries such as lithium iron phosphate/lithium half cells using cellulose/polysulfonamide

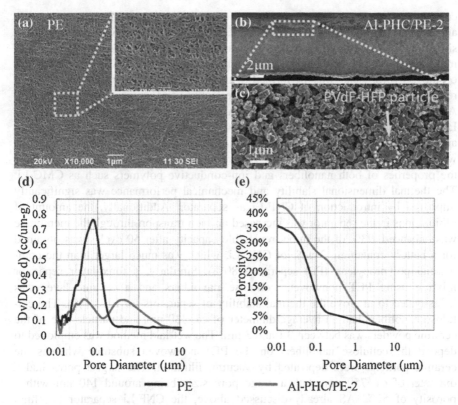

Fig. 2.14 The SEM images of: the PE (**a**); the fracture surface of the Al-PHC/PE-2 separator (**b**); the coated layer of Al₂O₃/PVdF-HFP (**c**). The pore distribution (**d**), and porosity curves (**e**) of the PE and Al-PHC/PE-2, respectively. Reproduced from Ref. [84] with permission of Elsevier

separators exhibited stable charge-discharge capability even at 120°. The authors proposed that these membranes provide promising new strategy for large-scale fabrication of high-performance lithium-ion battery membranes. A polydopamine coating layer and a cellulose/polydopamine (CPD) membrane was reported to produce membrane possessed a compact porous structure, superior mechanical strength and excellent thermal dimensional stability by a facile and cost-effective papermaking process [86]. It was shown that these membranes exhibited favorable properties (superior mechanical strength, low-cost and favorable electrochemical properties) for their use in LIBs. Lithium cobalt oxide/graphite cells having these separators CPD separator showed a good cycling stability and rate capability compared to commercially available polypropylene and neat cellulose separators. Furthermore, alternating-current impedance of this cell showed just minor variation of 9 Ω after the 100th cycle. Another approach to improve the performance of commercially available membranes was to coat polypropylene based separators [87]. This was accomplished by deposition of cellulosic aerogels based on hydroxyethyl cellulose (HEC), via ice segregation induced self-assembly. A cell consisting of the coated separator, Li foil as the counter and reference electrodes, and LiFePO$_4$ as the cathode was assembled. The performance in terms of dimensional stability, electrolyte uptake, ionic conductivity, cycling performance, was much better than its non-coated counterpart. An interesting detail is that the coating of the polypropylene was performed without the use of toxic solvents, which rendered the preparation process cost effective and environmentally benign. A composite membrane consisting of PVA and CNF-Li was proposed for LIBs by Liu and coworkers [88]. The membrane was prepared using the NIPS techniques and showed high porosity (>60%), good ionic conductivity (ca. 1.1 mS cm^{-1}) as well as remarkable electrolyte uptake. The main idea to use CNF-Li was to combine the properties of both nanofibers and ion-conductive polymers such as CMC-Li. The thermal dimensional stability and mechanical performance was significantly improved by introduction of CNF-Li as separator. Additionally, the amount of lithium ions in the separator was increased and ion transport through the membrane was enhanced. 93% of the initial reversible capacity after 50 cycles was obtained for a battery containing 2 wt% of CNF-Li, which was much larger than the commercial polypropylene (PP) separator (80%). Similarly, a three-layer separator mixture using PET nonwoven, cellulose nanofibers and a ceramic layer was investigated to improve the thermal stability of separators for LIBs [89]. The cellulose nanofibers had an average diameter of ca. 330 nm, and particle size of the ceramic powders was between 0.1 and 3 μm. The wet-laid method was employed to deposit the cellulose nanofibers on the PET nonwoven substrate whereas the ceramic particles were deposited by vacuum filtration. The largest pores had a diameter of ca. 750 nm, with average pore sizes being around 140 nm with a porosity of 52%. AS already discussed above, the CNF-Li separator has high affinity to the electrolyte (3.2), while shrinkage during thermal stress (160°/2 h) was not observed.

2.2.3 Non-woven Mats

A simple method to obtain non-woven materials is electrospinning. The resulting materials feature a high degree of porosity and feature rather small fiber diameters. The first example to be discussed in this context is the generation of PVDF/PMMA/cellulose acetate (CA) composite mats having different compositions [90]. A wide range of ratios (100:0:0, 90:10:0, 90:5:5 and 90:0:10) were successfully electrospun. Interestingly, CA favors the adsorption of electrolyte while the ratio of 90:0:10 yielded membranes with the highest porosity (99.1%) and electrolyte uptake (3.2). In another approach, a heat-resistant and flame-retardant cellulose-based composite nonwoven has been successfully employed as separator in LIBs (Fig. 2.15). The separator was based on pulp fibers, sodium alginate, a flame retardant agent and silica [91].

This separator featured a rather good flame retardant behavior and featured superior heat tolerance and proper mechanical strength. Further, electrolyte uptake

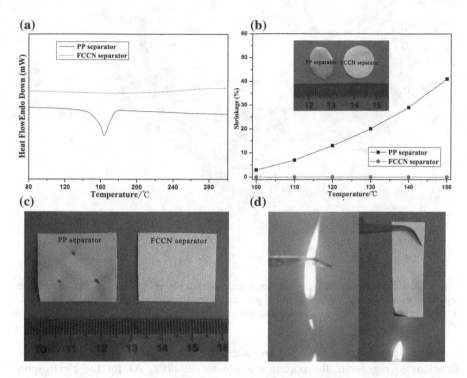

Fig. 2.15 a DSC curves of PP separator and FCCN separator. **b** Thermal shrinkage rate of PP separator and FCCN separator over a temperature range from 100 to 150 °C, and the inset is the photograph of PP separator and FCCN separator after thermal treatment at 150 °C for 0.5 h. **c** Contact test between hot electric iron tip and separators. **d** Combustion behavior of PP separator and FCCN separator. Reproduced from Ref. [91] under a Creative Common License 3.0

was increased and enhanced ionic conductivity was observed. A battery composed of lithium cobalt oxide/graphite using this separator had a better rate capability and cycling retention compared to a commercial PP separator. For a lithium iron phosphate/lithium cell having such an integrated composite separator, stable cycling performance and thermal dimensional stability up to 120 °C was observed. Zhang et al. [92] used a similar concept by electrospinning CA, followed by regeneration using LiOH and coating of the membrane with a 2 wt% solution of PVDF-HFP in acetone. The resulting separator showed good thermal stability and electrolyte uptake. Alcoutlabi and coworkers [93] compared two methods for the preparation of non-woven mats, namely electrospinning and forcespinning for potential use in LIBs. PVDF nanofiber coatings were deposited by electrospinning on polyolefin microporous membranes leading to improved electrolyte uptake and electrochemical performance. Forcespinning was employed to manufacture fibrous non-woven cellulose mats based on cellulose. Forcespinning is based on centrifugal forces which yield fine fibers. Compared to electrospinning, this technique is much less sensitive toward environmental influences (Fig. 2.16) [94].

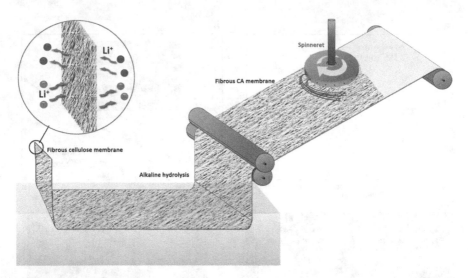

Fig. 2.16 Schematic representation of large-scale fabrication of fibrous cellulose membrane separator using the forcespinning technique and subsequent alkaline hydrolysis treatment. Reproduced from Ref. [94] with permission from Springer

For this purpose, CA was used and after processing it was treated with alkaline solutions to regenerate the porous scaffold to cellulose. As for the electrospun samples, the fibrous cellulose membrane-based separator exhibited high electrolyte uptake and good electrolyte/electrode wettability. CA from used cigarette filter was used by Huang and coworkers in the preparation of separators [95]. They manufactured a cellulose/PVDF-HFP nanofiber membrane by coaxial electrospinning of a cellulose acetate core and PVDF-HFP shell, followed by hydrolysis of the CA by LiOH. The resulting cellulose-core/PVD-HFP-shell fibrous membrane exhibited high tensile strength (34.1 MPa), high porosity (66%), reasonable thermal stability (up to 200°), and good electrolyte uptake (3.6). Additional features were the low interfacial resistance (98.5 Ω) and high ionic conductivity (6.16 mS cm^{-1}) compared to commercially used separators (280.0 Ω and 0.88 mS cm^{-1}). The rate capability (138 mAh g^{-1}) and cycling performance (75.4% after 100 cycles) were also better compare to commercial separators. Lalla et al. [96] used CNCs to reinforce electrospun PVDF-HFP copolymers. The content of CNCs in the mats was studied in dependence of the tensile strength, and the thermal properties, with 2 wt% of CNC being the optimum amount. At these filer content, tensile modulus increased by 75% over a temperature range from 30 to 150 °C. At higher CNC contents, the materials became more brittle and mechanical properties were inferior. Porous non-woven mats of MFC in combination with ceramics and PVDF have been used to generate separators for LIBs by Huang and coworkers [97]. The membranes showed good mechanical (strength of up to 10 MPa, Young's modulus of up to 658 MPa) and thermal stability (up to 180 °C). The ionic conductivity was good for a series of different membrane compositions with 1.28 mS cm^{-1} (in 1 M

LiPF$_6$/EC/DMC) being the best. Coin cells composed of LiNMC cathodes, LiPF$_6$ in EC/DMC 1:1 (v/v) electrolyte, and graphite anodes showed stable cycle performance.

2.2.4 Solid Polymer Electrolytes (SPE), Gel Polymer Electrolytes (GPE) and Composite Polymer Electrolytes

Solid polymer electrolytes (SPEs) are an interesting class of materials since they are potentially interesting for high energy density battery applications, electrochromic devices and sensors. Most of the described systems are based on polyethers. In such polymers, ion conduction is believed to proceed via local polymer chain relaxation which is favored for amorphous polymers having a low glass transition temperature. This means that for high ion conduction in SPE, amorphous matrices are required. However, for most of the systems described so far inherent conductivity limits have been reached (ca. 10^{-4} S cm^{-1}) which is far below the limits of most battery applications (ca. 10^{-3} S cm^{-1}). One particular case of SPE are gel polymer electrolytes (GPE). The incorporation of plasticizers (for increasing ion mobility by suppressing crystallization) in a crosslinked polymer matrix is capable to provide a mechanically stable matrix while retaining plasticizers and electrolyte. In this context, commonly used polymers are polyacrylonitrile, poly(vinylidene fluoride), poly(methyl methacrylate), poly(ethylene oxide) derivatives.

In recent years, biobased materials such as polysaccharides have gained the focus of research in the area of SPE/GPE since many polysaccharides readily form gels with low crystallinity. Particularly cellulose esters have been widely studied in this context. For instance, Yue and coworkers [98] studied a range of hydroxypropyl cellulose esters with oligomeric poly(oxyethylene) side chains. They employed lithium triflate as electrolyte and EC and PC as well as mixtures thereof as plasticizers. Conductivities in the range of 10^{-3} S cm^{-1} were reached at room temperature, when the plasticizer content was above 50%. For gels with higher plasticizer contents (60–70%), crosslinking using 1,6-diisocyanatohexane was performed. Depending on the amount of crosslinking agent either porous or non-porous films or gels were obtained. These materials showed conductivities in the range of 10^{-3} S cm^{-1} as well but had better mechanical properties compared to the non-crosslinked gels. In a similar way, hydroxyethyl cellulose and CM-cellulose can be grafted with PEO side chains [99].

Although the obtained materials are highly amorphous with T$_g$ below room temperature, the observed ionic conductivities in the presence of LiClO$_4$ were rather low, reaching values in the range of 10^{-6}–10^{-7} S cm^{-1}. In another report, HEC was plasticized with different amounts of glycerol and lithium triflate was added [100]. The resulting transparent films cast from water showed conductivities in the range of 10^{-5} S cm^{-1} at room temperature and 10^{-4} S cm^{-1} at 80 °C at a glycerol content of ca. 50%. The authors proposed this SPE for use in electrochromic devices but did not provide a proof of principle. A different approach was proposed by Ledwon and coworkers who cast HPC membranes from dichloromethane [101]. T_g points were detected at around −40 °C and degradation was observed in three stages starting at 130 °C. Ionic conductivities for the best samples ranged from 3.5×10^{-5} to 1.1×10^{-4} S cm^{-1} at 25 and 50 °C. These materials have been deposited on SnO$_2$: Sb/glass (ATO) electrodes and their performance was evaluated by cyclic voltammetry in the −2 to 1.5 V interval. Afterward, electrochromic devices were built using PEDOT:PSS and polyaniline/Prussian blue (PB) as electrochromic layers. Results from UV-Vis spectroscopy revealed 35% of color change at 650 nm for the device with glass/ITO/PB/HPC/PEDOT:PSS/ITO/glass configuration. After 20 cycles, the absorbance value just changed by 0.22.

HPC, sodium iodide, 1-methyl-3-propylimidazolium iodide (MPII, an ionic liquid), EC and PC were employed to prepare a non-volatile GPE for an envisaged use in dye-sensitized solar cell (DSSC) applications [102]. At a 1:1 ratio of MPII and HPC, the highest ionic cond. of 7.37×10^{-3} S cm^{-1} was achieved; however, even at lower ratios conductivities in the 10^{-3} S cm^{-1} regime were reported. The resulting materials are highly amorphous and T_gs are below −100 °C. These materials were then used in the manufacturing for DSSCs where the 1:1 MPII:HPC materials showed the best performance with energy conversion efficiency of 5.79%, with short-circuit current densities, open-circuit voltage and fill factor of 13.73 mA cm^{-2}, 610 mV and 69.1%, respectively.

Cyanoethylated hydroxypropyl cellulose (CN-HPC) was proposed as polymer gel electrolyte for application in DSSCs. The best samples exhibited rather good ionic conductivities in the 10^{-3} S cm^{-1} range using LiI/I$_2$ and 1-methyl-3-hexylimidazolium iodide (MHII)/I$_2$ as the I$^-$/I$_3^-$ redox couple with the respective diffusion constants. of I$_3^-$ (D app) of 2.54×10^{-6} cm^2 S^{-1}. Under the optimized condition, the overall conversion efficiencies of quasi-solid DSSCs were reported 7.40% using a triphenylamine dye (SD2) and 7.55% using a Ru dye (N719), which refers to 94% of the liquid electrolyte.

A similar approach was demonstrated by Sato and coworkers who used a semi IPN gel polymer, composed of CN-HPC multifunctional poly(oxyethylene) methacrylate in conjunction with LiCoO$_2$ covered by an ion conductive poly-urethane [103]. The ionic conductivity of this material was reported to be 2.7×10^{-3} S cm^{-1} at 25 °C. A rather large cell (2500 mAh) was assembled which showed good discharge performance and improved safety characteristics as proven by a nail penetration test. Furthermore, overcharging was prevented by this new battery system. The authors proposed their materials for applications in large batteries featuring inherent safety such as batteries for mobile applications.

Ren et al. [104] synthesized a gel polymer electrolyte by blending PVDF-HFP and HPC-CN) in DMF/glycerine solutions (6:1). After casting, rinsing and drying, free standing films were obtained. The conductivity of the neat PVDF-HFP was significantly improved (4.36 mS cm^{-1}) by the presence of the HPC-CN (14:1) and 1 M PF$_6$ in EC and DMC. For all compositions, the blend membranes were elec-trochemically stable up to about 4.8 V versus Li/Li$^+$. Ethyl cellulose has been used as gelator for the production of acetonitrile based electrolyte (LiI, I$_2$, *tert*-butyl-pyridine and tetrabutylammonium iodide) for dye solar cells [105]. This PGE had just slightly lower performance in terms of photovoltaic conversion efficiency (6.5% for liquid electrolyte vs. 5.9% for gel electrolyte) at an ethyl cellulose content of 5.8%. Interestingly, the electrolyte gelation had just minor effects on ionic diffusion coefficients of iodide, and devices were remarkably stable for at least 550 h under irradiation at 55°. An electrochromic device having an ethyl cellulose-based gel polymer electrolyte was created by Lin and coworkers [106]. The device consists of a MoO$_3$ film as the main cathodic electrochromic layer, a gel polymer electrolyte as an ion conduction layer, and a NiO film as the complementary, anodic electrochromic layer (Fig. 2.17).

The device featured good reversibility, low power consumption of −1.5 V in color state, high variation of transmittance (51.9%), changes in optical density (0.754), coloration efficiency of 54.9 cm^2/C and good memory effect under open-circuit conditions (Fig. 2.18).

It was demonstrated that ethyl cellulose-based gel polymer electrolytes were electrochemically stable; problems associated with electrolyte leakage were avoided. In a similar report, the use of ethyl cellulose and acid functionalized multi-walled C nanotubes (oMWCNTs) as a co-gelator and the application of these gels for quasi-solid state DSSCs was demonstrated [107]. The gels were prepared by blending ethyl cellulose and oMWCNTs with methoxypropionitrile (MPN) and acetonitrile, whereas ethyl cellulose and oMWCNTs contents can be in the range of 4 and 1.5%, respectively, to induce the gel formation. Without any oMWCNTs, the ethyl cellulose contents must be as high as 12% for the induction of gel formation. The PGEs were prepared by adding 1-methyl-3-propylimidazolium iodide (PMII), guanidinium thiocyanate and 4-CMe3 pyridine into the EC-oMWCNT/ACN-MPN gels. The obtained materials were then employed for the fabrication of quasi-solid state DSSCs. The photocurrent efficiency of the gel DSSCs was rather good with up 6.97% under AM1.5G illumination, which was in the same range as the liquid state DSSCs. However, the GPE DSSCs featured much higher stability (98% of initial PCE after 30 days vs. 80% of initial PCE).

Fig. 2.17 Schematic overview on the assembled device. Reproduced from Ref. [106] with permission from Elsevier

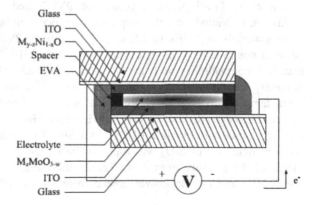

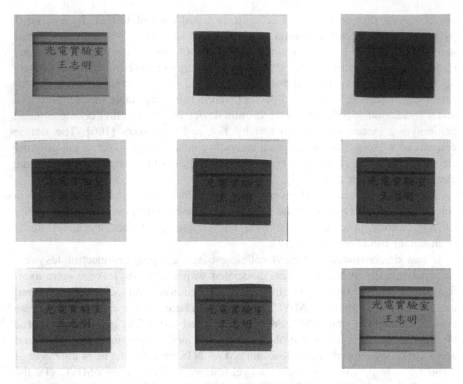

Fig. 2.18 The images of the sample 4(Glass/ITO/NiO/GPE/MoO$_3$/ITO/Glass) after various open-circuit times. Reproduced from Ref. [106] with permission from Elsevier

PGEs on the basis of linearly grafted chains of the lithium salt of 2-acrylamido-2-methylpropane sulfonic acid on ethyl cellulose have been prepared by using free radical initiator azobis(cyclohexanecarbonitrile) [108]. Afterwards, MMA and a crosslinking agent, ethylene glycol dimethacrylate, were added to form a semi-interpenetrating network. The resulting GPEs showed promising ionic conductivities in the 10^{-3} S cm^{-1} range.

Liu et al. [109] prepared blends of PVDF and cellulose acetate butyrate (CAB) on polyethylene (PE)-supported polymer membranes with the aim to use these materials as GPEs in LIBs. A 2:1 ratio of PVDF:CAB showed the best performance in terms of ionic conductivity (2.48×10^{-3} S cm^{-1}) while exhibiting excellent compatibility with the anode and cathode of the lithium ion battery. A battery based on LiCoO$_2$/graphite and this GPE showed good cyclic stability at room temperature, storage performance at elevated temperature and rate performance. Zhao extended this approach by incorporation of SiO$_2$ into the materials via a sol gel process to obtain a PVDF-CAB-SiO$_2$/PE blended GPE [110]. The optimum particle size of SiO$_2$ was found to be ca. 500 nm (at 5% loadings), which increased the porosity from 40 to 42.3%, the mechanical strength from 117.3 to 138.7 MPa and the electrolyte uptake from 149 to 195%. Further, the stability window opened from 4.7 to 5.2 V and the ionic conductivity at room temperature doubled (from 1.16 to 2.98 mS cm^{-1}).

A PGE base on a mixture of lithium bis(oxalato)borate (LiBOB), γ-butyrolactone (γ-BL), and CA was reported by Abiding et al. [111]. The PGE caused a significant increase in ion conductivity and values of up to 7.05 mS cm^{-1} at 2.4 wt% CA were determined and the plots of conductivity versus temperature showed a classical Arrhenius relationship. The authors looked into the mechanism of the ionic conduction in their PGE and found that the material followed the SPH model. This model implies that a small polaron is formed by the addition of a charge carrier to a site that presents a network of sites for the ions to reside. The activation energy for hopping is the same as the activation energy for conduction meaning that the energy required for lithium ions to jump to the adjacent site is the same for conducting and relaxing.

Solid polymer electrolytes (SPEs) were also prepared by blending LiClO$_4$, methyl cellulose (MC) and an oligometric dendritic polyethylene glycol (PEG), formed from polyoctahedralsilsesquioxane functionalized with ~ 8 PEG side chains (POSS-PEG) on the SiO$_{1.5}$ core [112]. Three different compositions (80/20, 70/30 and 60/40 POSS-PEG/MC), and LiClO$_4$ (O/Li = 16/1, corresponds to POSS-O to Li$^+$) were investigated towards their electrochemical, mechanical and morphological behavior. A major advantage of this system is the tenability of the materials from rather hard to elastic by changing the ratio between MC and POSS-PEG. As known for other polymers, the POSS-PEG-LiClO$_4$ was microphase separated from the MC phase as shown by SEM and TEM. The POSS-PEG-LiClO$_4$ phase in the blends was highly amorphous between -100 °C and its decomposition temperature (ca. 300 °C), while the MC phase shows semicrystalline behavior proven by XRD. Interestingly, the moduli of the blends increased with increasing POSS-PEG content below the glass transition temperature which indicated that the POSS core acted like a reinforcing agent. At higher temperatures, the moduli of the blends increased with MC content, which was related to crosslinking of OH groups of MC with the dominant POSS-PEG/LiClO$_4$ phase (32–156 MPa). Ionic conductivities of 1.6×10^{-5} and 1.1×10^{-6} S/cm were determined at 30 and 0 °C, respectively, for the 80/20 POSS-PEG/LiClO$_4$ (O/Li = 16/1)/MC blend. Stability and reversibility of the blends at 50 °C were observed in the range 1.5–4.2 V.

Composite polymer gels were obtained from CA, N-methyl-N-propylpyrrolidinium bis(trifluoromethanesulfonyl)imide (PyrTFSI), and lithium bis (trifluoro-methanesulfonyl)imide (LiTFSI) [113]. The resulting ionic gel formed a completely homogeneous phase at the molar ratio of 1:3:1.5 as shown by DSC whereas the ionic conductivity of the PGEs was significantly enhanced by the presence of LiTFSI. This probably originates from a strong interaction of the Li$^+$ with the carbonyl group of CA as indicated by FTIR spectroscopy.

Electrospinning of PVDF/CA blends for GPE in LIBs was investigated by Kang et al. [114]. The GPE with a CA:PVdF = 2:8 ratio (in wt.) significantly improved the basic parameters such as strength (11.1 MPa), electrolyte uptake (7.7), thermal stability (no shrinkage under 80 °C without tension), and ionic conductivity (2.61×10^{-3} S cm^{-1}). A Li/GPE/LiCoO$_2$ battery was assembled which showed good cyclic stability and storage performance at room temperature, reaching a specific capacity up to 204 mAh g^{-1}.

Composite nanofiber membranes based on CA/poly-L-lactic acid (PLLA)/halloysite nanotube (HNT) were employed for GPEs in LIBs. The crystallization behavior of the polymeric materials was significantly suppressed while thermal stability was improved by the incorporation of HNTs and an ionic conductivity of 1.52×10^{-3} S cm^{-1} was obtained. The authors compared their GPE to Celgard 2500 in terms of performance of a Li/GPE/LiCoO$_2$ and judged their new system as more competitive.

Another more application oriented study reported optimized conditions for the use of CA in GPEs for electrochromic devices (ECDs) [115]. Lithium perchlorate (LiClO$_4$) was added as a supporting electrolyte and propylene carbonate as a solvent to the GPE. Poly(3-hexylthiophene-2,5-diyl) (P3HT) thin films were produced to act as an electrochromic cathodic layer on an electrochromic anodic ITO substrate with the GPE being the ion conducting material in between (Fig. 2.19). The stability of the P3HT film by measuring the optical and electrochemical properties of P3HT thin films on ITO in GEs through UV-Vis and cyclic voltammograms obtained during application of potential to the films.

Current/voltage data and related performance for ECDs were investigated and revealed that the GPEs featured high stability within the operative potential window for ECDs. The electrochromic polymer films showed also fully reversible color change for more than 1000 cycles without polymer film or GPE degradation. This work shows the first example of the use of GE with a natural polymer matrix in electrochromic devices and demonstrates their reliability under repetitive switching of applied voltage for up to 1000 cycles.

Fig. 2.19 Electrochromic device colored state and bleached state with the application of +1.0 V$_{dc}$ (*dark purple* to *transparent blue*) and −1.0 V$_{dc}$ (*transparent blue* to *dark purple*). Reproduced from Ref. [115] with permission of Springer

1st cycle

1000th cycle

Butyl-*N*-methyl pyrrolidinium bis(trifluoromethylsulfonyl) imide (PYR14TFSI) and methyl cellulose have been employed to produce a new type of mechanically robust, solid polymer electrolyte ion gels [116], with moduli in the MPa range, a capacitance of 2 μF cm^{-2}, and good ionic conductivities ($>1 \times 10^{-3}$ S cm^{-1}) at room temperature. The preparation of the gels was performed by dissolution of PYR14TFSI and MC in DMF followed by a heating/cooling cycle. After evaporation of the DMF, a thin, flexible, self-standing ion gel with up to 97 wt% PYR14TFSI was obtained which showed excellent conductivity and a large electrochemical operating window (5.6 V).

A composite gel polymer electrolyte on the basis of a nonwoven fabric (NWF) and methyl cellulose, prepared by a simple casting process, followed by soaking with electrolyte, was described by Li and coworkers [117]. The obtained materials featured good mechanical properties and good thermal and electrochemical stability due to the synergistic action between the methyl cellulose matrix and the NWF framework. For instance, the composite gel polymer electrolyte exhibited higher ionic conductivity (0.29 mS cm^{-1}) at room temperature and better lithium ion transference number (0.34) than those of the conventional Celgard 2730 separator (0.21 mS cm^{-1} and 0.27) in 1 M $LiPF_6$ electrolyte. Cells ($Li/LiFePO_4$) using this composite gel membrane exhibited better cycling retention and higher discharge capacity than those based on Celgard 2730 separator and pure MC gel membrane.

In another study, CNF was used to synthesize four composite electrolytes for lithium ion battery applications [118]. The GPE was composed of a ionically conductive PEG matrix reinforced with nanofibrillated cellulose (CNF) to provide mechanical integrity. In order to achieve good compatibility between the CNFs and the PEG, a propionate and acrylate based modification was carried out to enable the formation of covalent bonds between the PEG and cellulose phase. This was particularly beneficial for the uptake of liquid electrolyte to enhance the ionic conductivity. Although this was a very interesting approach, the ionic conductivities (5×10^{-5} S cm^{-1}) could not fully compete with the best available GPEs based on polysaccharides. Similarly also MFC can be used to improve material characteristic of GPEs. A fully-solid methacrylic-based thermo-set polymer electrolyte membrane reinforced with micro-fibrillated cellulose (MFC) was described by Chiappone [119]. The membrane was manufactured in water and crosslinking was initiated by UV-induced polymerization via a free radical mechanism. The synthesized GPE exhibited excellent mechanical properties with a Young's modulus as high as 32 MPa while ionic conductivity was still acceptable (0.1 mS cm^{-1} at 50 °C). Similar work has also been reported by several other authors, with the main aim to improve compatibility and mechanical performance. In most cases, methacrylic membranes were photochemically crosslinked with cellulosic materials (hand sheets, papers, membranes, microparticles) using UV grafting [120]. In most cases, mechanical properties were significantly improved and in some cases electrochemically competitive materials were obtained [121]. One of the main drawbacks that restrict the practical application of gel-polymer electrolytes is the inferior mechanical performance compared to other available systems.

Electrochromic devices were investigated by Kiristi who used WO_3 (cathodically coloring working electrode) and NiO (anodically coloring counter electrode) in conjunction with GPEs based on CMC [122]. The performance evaluations of the complementary solid-state electrochromic devices indicated good reversibility, low power consumption of ± 3 V in colored state, high variation of transmittance changing of 64% in gel electrolyte, and lower 7% in membrane electrolyte and good memory effect under open-circuit conditions.

CMC membranes have been used as a host for a GPE in lithium ion batteries [123]. The morphology of the membrane can be fine-tuned by variation of fine-adjusted by varying the ratio of the solvent and non-solvent mixture. The membrane was capable to take up 76% electrolyte whereas the ionic conductivity (1 M $LiPF_6$) at room temperatures reached up to 0.48 mS cm^{-1}. Lithium ion transference was reported 0.46 (compare Celgard: 0.27) and when a battery was assembled ($LiFePO_4$), the materials exhibited excellent electrochemical performance (higher reversible capacity, better rate capability, good cycling behavior). In another report, the motivation was to spread the electrochemical operating window of aqueous 2 M Li_2SO_4 and 4 M $LiClO_4$ by addition of GPEs [124]. The main result was that in the case of Li_2SO_4-based electrolytes, the addition of CMC or agarose increased the stability window.

PVA/CNC/SiO_2 nanocomposite films were prepared in different ratios as SPE for fuel cells. The impact of CNCs on the performance was evaluated in a range from 20 to 60 wt% CNC content. Realized conductivities ranged from 0.044 to 0.065 S cm^{-1} at 20 and 60 °C, respectively. The CNCs improved the dimensional stability while maintaining the conductivity of existing anion exchange membranes [125].

A PEO-CMC GPE for DSSCs was reported by Bella et al [126]. The GPE was prepared by soaking liquid electrolyte containing supporting salts and I_3^-/I^- redox couple into the polymer blend. A photovoltaic-chemometric approach allowed for assembling a device with efficiencies up to 5.18% under 1 sun irradiation ($\sim 7\%$ under 0.4 sun). The durability of the device was good, showing excellent efficiency as high as 98% even after 250 h under extreme aging conditions.

A similar application was also reported for agarose based gels [127]. GPEs with different agarose concentrations (1–5 wt%) and various inorganic filler concentrations (0–10 wt% TiO_2) were investigated. By increasing the agarose and inorganic filler amount, a decrease in T_g in the range of 1–2 wt% for agarose and 0–2.5 wt% for TiO_2 was observed, which results in high conductivity. The electron lifetime in TiO_2 of DSSCs increased with agarose contents, while it decreased with inorganic filler contents. A cell with the electrolyte of 2 wt% agarose revealed an optimum energy conversion efficiency of 4.1% while optimum efficiency of the DSSC with added TiO_2 is 4.74% at 2.5 wt% loading.

A composite GPE based on thermoplastic polyurethane (TPU) and cellulose was fabricated and studied for applications in Li batteries [128]. The GPE featured acceptable ionic conductivities (4.8×10^{-4} S cm^{-1}) at 80°, high Li ion transport number (0.68) and improved electrochemical stability. An assembled $LiFePO_4$/Li battery using the TPU/cellulose GPE exhibited good rate capacity and remarkable cycle performance at 80°. The discharge capacity was still 128.2 mAh g^{-1} after 200 cycles, 95% of the capacity retention at a charge/discharge rate of 2 C.

PEO nano-fiber mats with MCC/CNC have been reported by Samad and coworkers [129] for envisaged use in SPE is DSSCs. The reinforced fiber mats showed a two-fold increase in the tensile strength and up to five times increase in the Young's modulus. Thermal behavior studies revealed that the electrolyte was stable up to temperature of 200 °C. Ionic conductivity in the order of 10^{-4} S cm^{-1} was achieved for the samples at 100 °C, which is at the lower end for applications in LIBs. In a different approach, non-modified MCC was used in combination with different ionic liquids (MPII, EMISCN) to form GPE for DSSCs [130]. The photovoltaic performance of cellulose gel-based DSSCs has been optimized by monitoring some key parameters, such as ionic liquid volume ratios and cellulose contents. PCEs of up to 3.3% have been realized without any organic solvents, and good stability was demonstrated during 8 h of exposition to simulated solar light. A similar approach was chosen by Li et al. [131] with the main difference that cellulose was grafted with acrylic acid in 1-butyl-3-methylimidazolium iodide ([Bmim]I) as reaction medium. KI and I$_2$ were chosen as ionic conductors which were already dispersed prior to polymerization in BMIMI, and a conductivity of up to 7.33 mS cm^{-1} was realized. For the optimized composition of the GPE, a DSSC was assembled with a PCE of 5.51% at 100 mW cm^{-2}.

Navarra and coworkers synthesized cellulose-based hydrogels (from different cellulose sources) via low-cost synthetic routes for GPE membranes [132]. A crosslinking step was introduced to tune liquid uptake capability and ionic conductivity. For this purpose, the redox behavior of electroactive species (Fe(CN)$_6^{4-}$/Fe(CN)$_6^{3-}$) entrapped into the hydrogels has been investigated by cyclic voltammetry tests, revealing very high reversibility and ion diffusivity.

The same redox couple (Fe(CN)$_6^{4-}$/Fe(CN)$_6^{3-}$) as in the previous report was employed to realize thermoelectrochemical cells [133] (TECs). TECs are a promising and cost-effective approach to harvesting waste thermal energy. The electrolyte with 5 wt% cellulose achieved an optimum balance of mechanical properties, Seebeck and diffusion coefficients and supported power outputs comparable to those of the liquid electrolyte systems.

GPEs for supercapacitors have been realized using different types of cellulose-chitin hybrid gels. In all these applications ionic liquids (e.g. 1-butyl-3-methylimidazolium, 1-allyl-3-methylimidazolium bromide, 1-butyl-3-methylimidazolium chloride) are used. In all these applications, electric double layer capacitors (EDLC) were reported [134]. Test cells with a hybrid gel electrolyte showed a specific capacitance of 162 F g^{-1} at room temperature, which was higher than that for a cell with an H$_2$SO$_4$ electrolyte (155 F g^{-1}). The discharge capacitance of the test cell was able to retain over 80% of its initial value in 100,000 cycles even at a high current density of 5000 mA g^{-1}. Ionic conductivity reached up to 57.8 S m^{-1} at room temperature. The self-discharge measurements suggested that leakage current and potential decay were suppressed by the application of the acidic cellulose-chitin hybrid gel electrolyte. These results indicated that the acidic cellulose-chitin hybrid gel electrolyte had a practical applicability to an advanced

EDLC with excellent stability and working performance. A similar concept can also be used for other polysaccharides. The same authors used alginates and chitosan for the realization of EDLCs [135]. As ILs, hydrophobic 1-ethyl-3-methylimidazolim tetrafluoroborate (EMImBF4) was used resulting in mechanically strong gels featuring a high retention of EMImBF$_4$. According to charge/discharge measurements, the EDLC with alginate- and chitosan-based gel electrolytes exhibited excellent discharge capacitance. Probably, the alginate featured high affinity to the activated carbon electrode leading to a decrease in the electrode/electrolyte interfacial resistance. A test cell with Alg/EMImBF$_4$ did not show any decrease of coulombic efficiency (99.8%) during 5000 cycles.

A nanocomposite consisting of PEO/LiClO$_4$/chitin nanocrystals (ChNC), prepared by hot pressing, was proposed as PGE [136]. The ChNC acted as reinforcing agent to improve mechanical properties of the PEO and the ionic conductivity of the GPE was increased by one order of magnitude and the lithium transference number, t_{Li+}, rose from 0.24 to 0.51.

Chitosan and its derivatives have also been studied for SPE/GPE. Different types of chitosan acetate films have been employed whereas the dependency of the amount of plasticizer on the ionic conductivity was investigated [137]. Films of chitosan acetate, plasticized chitosan acetate, chitosan acetate containing electrolyte, and plasticized chitosan acetate-electrolyte complexes were cast. It turned out that all films were highly amorphous favoring ion transport throughout the GPE.

Pectin-based gel electrolytes in a transparent film form were obtained by a plasticization process with glycerol and addition of LiClO$_4$ [138]. The ionic

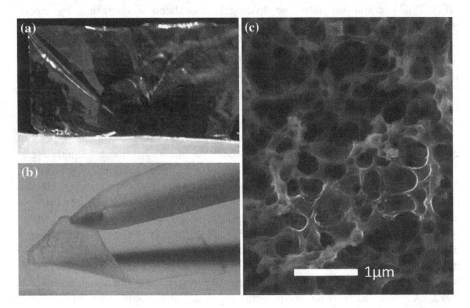

Fig. 2.20 a KGM thin film. **b** KGM thin film after adsorb electrolyte on substrate bended by tweezer. **c** SEM image of KGM thin film. Reproduced from Ref. [139] with permission of Elsevier

conductivity of the films was in an acceptable range whereas the glycerol content directly correlated with the conductivity at room temperature (from 10^{-5} for 37% glycerol to 5×10^{-4} S cm^{-1} for 68% glycerol). The authors proposed the materials for usage as gel electrolytes in electrochromic devices.

Konjac glucomannan (KGM) was proposed as GPE in CdS/CdSe quantum dot-sensitized solar cells (Fig. 2.20) [139]. The KGM was directly applied on a Cu_2S counter electrode in one step without mold, which simplified the cell fabrication process. The conductivity of the KGM was determined to be 0.074 S cm^{-1} at room temperature. The cell based on this GPE exhibited a PCE of 4.0% at 100 mW cm^{-2} with excellent stability compared to that of liquid-based quantum dot solar cells.

2.3 Electrode Materials from Polysaccharides

In the past decade, a wide range of polysaccharide based sources has been employed for the generation of carbonaceous materials for electrode fabrication. In order to get porous structures usually a pre-carbonization step is employed followed by carbonization using different types of additives. These additives either create highly porous structures (e.g. KOH) having high specific surface area or are used to incorporate additional functionality for doping purposes (e.g. N or P). The applications of such materials are manifold and cover all areas of energy storage and conversion with supercapacitors being currently the most important one. A wide range of materials has been used as carbon precursor for supercapacitors and an exemplary list containing just waste materials is depicted in Table 2.1.

However, supercapacitors based on polysaccharides have been covered in a separate book and will not be covered explicitly again here. We refer the reader to a recently published book [155]. Further, for batteries high specific surface area is not necessarily an advantage since electrochemical processes in the bulk are most important.

In the following, a few selected examples will be provided which strategies can be followed for successfully implementing polysaccharide derived carbons into battery applications. One strategy is to use the polysaccharide as template during carbonization to control the aggregation and/or clustering of additional electrode materials. A very convenient example is the use of chitosan which was used to control the formation of citrate capped iron oxide composites which subsequently were employed as anode materials in LIBs [156]. In the first step, the chitosan effectively interacted with the citrate capped Fe_3O_4 (C–Fe_3O_4) nanoparticles via electrostatic interactions between carboxylate groups of C–Fe_3O_4 and free amine groups of the chitosan. The second step, calcination of chitosan–linked Fe_3O_4 particles, led to carbon–coated Fe_2O_3 (Fe_2O_3@carbon) with a high degree of mesoporosity (pore size: 20–30 nm). This mesoporous Fe_2O_3@carbon composite exhibited a rather high capacity retention which was twice of bare Fe_2O_3 after the 50th cycle at 0.1 C. The amount of crosslinking of the iron oxides was studied and

Table 2.1 Waste materials used in the preparation of supercapacitors and their corresponding electrochemical performance parameters

Carbon source	Pyrolysis (°C)	SSA (m² g⁻¹)	C_m (F g⁻¹)	Electrolyte	Capacitance retention
Potato waste [140]	700	1052	255 at 1 A g⁻¹	2 M KOH	93.7% after 5000 cycles at 5 A g⁻¹
Rice brans [141]	700	2475	265 at 10 A g⁻¹	6 M KOH	87% after 10,000 cycles at 10 A g⁻¹
Coconut shell [142]	800	2440	246 at 0.25 A g⁻¹	0.5 M H₂SO₄	93% after 2000 cycles at 0.25 A g⁻¹
Corn husks [143]	800	928	356 at 1 A g⁻¹	6 M KOH	95% after 2500 cycles at 5 A g⁻¹
Bamboo [144]	750	169	171 at 1 A g⁻¹ 221 at 1 A g⁻¹	1 M KOH 1 M H₂SO₄	92%/90% after 2000 cycles at 4 A g⁻¹
Fish scale [145]	700	1300	332 at 1 A g⁻¹	6 M KOH	100% after 5000 cycles at 1 A g⁻¹
Hemp [146]	1000	1173	204 at 1 A g⁻¹	1 M LiOH	99% after 10,000 cycles at 10 A g⁻¹
Willow catkin [147]	HT, MnO₂	234	189 at 1 A g⁻¹	1 M Na₂SO₄	98.6 after 1000 cycles at 1.0 A g⁻¹
Cabbage [148]	800	3102	336 at 1 A g⁻¹	2 M KOH	95%/after 2000 cycles at 5 A g⁻¹
Soybean curd residue [149]	700	582	215 at 0.5 A g⁻¹	2 M KOH	92% after 5000 cycles at 5 A g⁻¹
Cat tail [150]	850	1951	336 at 2 mV s⁻¹	6 M KOH	Not given
Pomelo peel [151]	600	2105	342 at 1 A g⁻¹	2 M KOH	Not given
Banana peel [152]	1000	1650	206 at 1 A g⁻¹	6 M KOH	98.3 after 1000 cycles at 10 A g⁻¹
Sunflower seed shell [153]	700	2585	311 at 0.25 A g⁻¹	30% KOH	Not given
Coffee beans [154]	900	1840	361 at 1.0 A g⁻¹	1 M H₂SO₄	95% after 10,000 cycles at 5 A g⁻¹

SSA denotes to the specific surface area and C_m to the mass specific capacitance

it was found that the capacity increased with increasing chitosan volume. Porosity can also be introduced by carbonizing already porous cellulosic materials such as aerogels. Huang and coworkers reported on a three-dimensional (3D) carbonaceous aerogel which was obtained from carbonization of bacterial cellulose (BC) [157]. The 3D carbonized BC (CBC) had a highly interconnected nanofibrous structure and featured rather good electric conductivity as well as mechanical stability and was proposed as electrode in Li–S batteries. The intrinsic porous structure allowed for a relatively high sulfur loading of 81 wt%. The authors demonstrated that the sulfur species were well dispersed and wrapped around the CBC nanofibers. Although the loading as already high, the S/CBC composite still contained free space to accommodate the volume expansion of sulfur during lithiation. A further advantage was that an ultralight CBC interlayer, placed in between the sulfur cathode and separator, resulted in a significant improvement in active material utilization, cycling stability, and Coulombic efficiency. The authors also indicated that, to some extent, the CBC interlayer was capable to absorb migrating poly-sulfides. The nanofibrous nature of the CBC interlayer acted as an additional col-lector for sulfur and thus could prevent the over-aggregation of insulated sulfur on the cathode surface. Another CBC material was reported by Wang et al. [158]. They showed that the CBC materials could be used in LIBs using 1 M PF6 in DMC/EC electrolytes using PVDF as binders. The electrochemical performance of the materials was very good, exhibiting high capacities (386 mAh g^{-1} at 0.2 C over 100 cycles) and stable cycling behavior with low capacity fading (ca. 0.07% per cycle). CBC involving Fe_2O_3 for use in LIBs have been reported by Huang [159]. The materials were simply prepared by soaking BC sheets with a solution containing iron nitrate, followed by thermal treatment. As for the previous reports, superior electrochemical performance was reported compared to neat carbon materials.

Carbonized filter paper was employed for preparation of electrode materials involving Sn@C nanospheres as negative electrode materials for SIBs [160]. Both, the carbonization of the filter paper and the reduction of SnO_2 to elemental Sn proceeded at the same time. During the carbonization, the Sn aggregated to form nanoparticles which were encapsulated with carbon sheaths. It was proposed by the authors that the formation mechanism of the spherical Sn@C nanospheres was based on the unique properties of molten tin which preferably forms with spherical droplets. Slices of the carbonized filter paper decorated with Sn@C nanospheres were assembled into electrochemical cells without any further treatment or additives and higher capacities and better Coulombic efficiencies than that of bare carbonized filter paper were observed.

In a different approach, the cellulose based material was used in paper based electrodes [161]. In the first step, LTO, cellulose nanofibers (C-CNF) and CNTs were processed in a papermaking process (Fig. 2.21). The obtained sheet-like materials were subjected to carbonization and hierarchical nanocomposites were obtained which served as flexible free-standing paper anode and as lightweight current collector for lithium-ion batteries at the same time. The in situ carbonization of CNF/CNT hybrid film immobilized with uniform-dispersed LTO induced a large increase in electrical conductivity and specific surface area. Therefore, the car-bonized paper anode exhibited extraordinary rate and cycling performance com-pared to the paper anode without carbonization. Another option to create high performance composites is to blend CNFs with well aligned GO [162]. After car-bonization, the obtained microfibers featured a conductivity of 649 ± 60 S/cm whereas the GO acted as a template during CNF carbonization. Further, the car-bonized CNF can accomplish for defects of reduced GO (rGO) and linked rGO

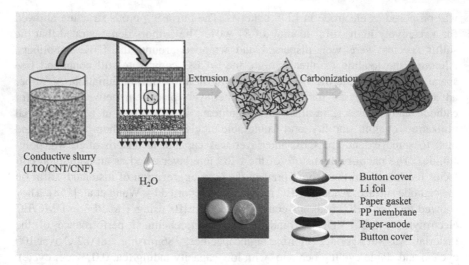

Fig. 2.21 Schematic illustration of preparation process of the flexible single-layer paper electrode. Reproduced from Ref. [163] with permission of the American Chemical Society

sheets together. The conductive microfibers showed promising behavior for a use in LIBs.

In another work, ordered cellulose nanocrystals (CNCs) have been transformed into porous carbon [163]. The obtained carbonaceous materials were characterized by an increased short-range ordered lattice and percolated carbon nanofiber at a carbonization temperature of 1000 °C. The CNC derived porous carbons showed superior performance with one of the highest reversible capacities (340 mAh g^{-1} at 100 mA g^{-1}) so far reported for carbon anodes for SIBs. The excellent electrochemical performance (rate capability and cycling stability) was correlated to the larger interlayer spacing, porous structure, and high electrical conductivity arising from the ordered carbon lattice and the percolated carbon nanofiber. This was supported by both molecular dynamic simulations and in situ TEM measurements.

In a different study, pulp fibers were used as precursor for carbons with intended use in SIBs [164]. The fibers have been pretreated with TEMPO before they were subjected to carbonization and electrode manufacturing. The obtained materials showed good performance as anode material with Coulomb efficiency of 72%, and a capacity of 240 mAh g^{-1} over more than 200 cycles. Zhu et al. [165] prepared an anode material which consisted of a Sn thin film deposited on a hierarchical pulp fiber substrate. The authors proposed that the 'soft' nature of the pulp fibers is capable to compensate for the mechanical stresses associated with the sodiation process, while its the mesoporous structure acted as an electrolyte reservoir, allowing for ion transport through the outer and inner surface of the fiber. An initial capacity of 339 mAh g^{-1} was reported and stable cycling over 400 cycles was observed.

References

1. Dash R, Pannala S (2016) Sci Rep 6:27449
2. Li J, Lewis RB, Dahn JR (2007) Electrochem Solid-State Lett 10:A17
3. Lestriez B, Bahri S, Sandu I, Roué L, Guyomard D (2007) Electrochem Commun 9:2801
4. Pejovnik S, Dominko R, Bele M, Gaberscek M, Jamnik J (2008) J Power Sour 184:593
5. (a) Hochgatterer NS, Schweiger MR, Koller S, Raimann PR, Wöhrle T, Wurm C, Winter M (2008) Electrochem Solid-State Lett 11:A76; (b) Bridel JS, Azais T, Morcrette M, Tarascon JM, Larcher D (2011) J Electrochem Soc 158:A750
6. Mazouzi D, Lestriez B, Roué L, Guyomard D (2009) Electrochem Solid-State Lett 12:A215
7. Tranchot A, Idrissi H, Thivel PX, Roue L (2016) J Electrochem Soc 163:A1020
8. Delpuech N, Mazouzi D, Dupré N, Moreau P, Cerbelaud M, Bridel JS, Badot JC, De Vito E, Guyomard D, Lestriez B, Humbert B (2014) J Phys Chem C 118:17318
9. Bridel JS, Azaïs T, Morcrette M, Tarascon JM, Larcher D (2010) Chem Mater 22:1229
10. Casalegno M, Castiglione F, Passarello M, Mele A, Passerini S, Raos G (1804) Chemsuschem 2016:9
11. Maver U, Znidarsic A, Gaberscek M (2011) J Mater Chem 21:4071
12. Cerbelaud M, Lestriez B, Guyomard D, Videcoq A, Ferrando R (2012) Langmuir 28:10713
13. (a) Oumellal Y, Delpuech N, Mazouzi D, Dupre N, Gaubicher J, Moreau P, Soudan P, Lestriez B, Guyomard D (2011) J Mater Chem 21:6201; (b) Mazouzi D, Delpuech N, Oumellal Y, Gauthier M, Cerbelaud M, Gaubicher J, Dupré N, Moreau P, Guyomard D, Roué L, Lestriez B (2012) J Power Sour 220:180; (c) Radvanyi E, Porcher W, De Vito E, Montani A, Franger S, Larbi, SJS (2014) Phys Chem Chem Phys 16:17142
14. Jeschull F, Lindgren F, Lacey MJ, Björefors F, Edström K, Brandell D (2016) J Power Sour 325:513
15. Lee J-H, Choi Y-M, Paik U, Park J-G (2006) J Electroceram 17:657
16. Drofenik J, Gaberscek M, Dominko R, Poulsen FW, Mogensen M, Pejovnik S, Jamnik J (2003) Electrochim Acta 48:883
17. Courtel FM, Niketic S, Duguay D, Abu-Lebdeh Y, Davidson IJ (2011) J Power Sour 196:2128
18. Etiemble A, Tranchot A, Douillard T, Idrissi H, Maire E, Rouee L (2016) J Electrochem Soc 163:A1550
19. Li J, Klöpsch R, Nowak S, Kunze M, Winter M, Passerini S (2011) J Power Sour 196:7687
20. Kim GT, Jeong SS, Joost M, Rocca E, Winter M, Passerini S, Balducci A (2011) J Power Sour 196:2187
21. Li J, Wang J, Wexler D, Shi D, Liang J, Liu H, Xiong S, Qian Y (2013) J Mater Chem A 1:15292
22. Courtel FM, Duncan H, Abu-Lebdeh Y, Davidson IJ (2011) J Mater Chem 21:10206
23. Lavoie N, Malenfant PRL, Courtel FM, Abu-Lebdeh Y, Davidson IJ (2012) J Power Sour 213:249
24. Zhang R, Yang X, Zhang D, Qiu H, Fu Q, Na H, Guo Z, Du F, Chen G, Wei Y (2015) J Power Sour 285:227
25. Yeo J-S, Yoo E-J, Ha S-H, Cheong D-I, Cho S-B (2016) J Power Sour 313:91
26. Hong X, Jin J, Wen Z, Zhang S, Wang Q, Shen C, Rui K (2016) J Power Sour 324:455
27. Bao W, Zhang Z, Gan Y, Wang X, Lia J (2013) J Energy Chem 22:790
28. Wei L, Chen C, Hou Z, Wei H (2016) Sci Rep 6:19583
29. Koo B, Kim H, Cho Y, Lee KT, Choi N-S, Cho J (2012) Angew Chem Int Ed 51:8762
30. Xu J, Chou S-L, Gu Q-F, Liu H-K, Dou S-X (2013) J Power Sour 225:172
31. Mancini M, Nobili F, Tossici R, Marassi R (2012) Electrochim Acta 85:566
32. Mancini M, Nobili F, Tossici R, Wohlfahrt-Mehrens M, Marassi R (2011) J Power Sour 196:9665
33. Zhang Z, Zeng T, Lai Y, Jia M, Li J (2014) J Power Sour 247:1

34. Zaïdi W, Oumellal Y, Bonnet JP, Zhang J, Cuevas F, Latroche M, Bobet JL, Aymard L (2011) J Power Sour 196:2854
35. Sun M, Zhong H, Jiao S, Shao H, Zhang L (2014) Electrochim Acta 127:239
36. Yue L, Zhang L, Zhong H (2014) J Power Sour 247:327
37. He J, Wang J, Zhong H, Ding J, Zhang L (2015) Electrochim Acta 182:900
38. Xie L, Zhao L, Wan J-L, Shao Z-Q, Wang F-J, Lv S-Y (2012) J Electrochem Soc 159:A499
39. Dahbi M, Nakano T, Yabuuchi N, Ishikawa T, Kubota K, Fukunishi M, Shibahara S, Son J-Y, Cui Y-T, Oji H, Komaba S (2014) Electrochem Commun 44:66
40. Sen UK, Mitra S (2013) ACS Appl Mater Interfaces 5:1240
41. Wang Z, Madhavi S, Lou XW (2012) J Phys Chem C 116:12508
42. Chou S-L, Gao X-W, Wang J-Z, Wexler D, Wang Z-X, Chen L-Q, Liu H-K (2011) Dalton Trans 40:12801
43. Zhong H, Zhou P, Yue L, Tang D, Zhang L (2014) J Appl Electrochem 44:45
44. Carvalho VD, Loeffler N, Kim G-T, Marinaro M, Wohlfahrt-Mehrens M, Passerini S (2016) Polymers 8:276
45. Lee B-R, Kim S-J, Oh E-S (2014) J Electrochem Soc 161:A2128
46. Cuesta N, Ramos A, Cameán I, Antuña C, García AB (2015) Electrochim Acta 155:140
47. Chen D, Yi R, Chen S, Xu T, Gordin ML, Wang D (2014) Solid State Ionics 254:65
48. Jeong YK, Kwon T-W, Lee I, Kim T-S, Coskun A, Choi JW (2015) Energy Environ Sci 8:1224
49. Ling M, Xu Y, Zhao H, Gu X, Qiu J, Li S, Wu M, Song X, Yan C, Liu G, Zhang S (2015) Nano Energy 12:178
50. Li G, Ling M, Ye Y, Li Z, Guo J, Yao Y, Zhu J, Lin Z, Zhang S (2015) Adv Energy Mater 5 doi:10.1002/aenm.201500878
51. Kuruba R, Datta MK, Damodaran K, Jampani PH, Gattu B, Patel PP, Shanthi PM, Damle S, Kumta PN (2015) J Power Sour 298:331
52. Kovalenko I, Zdyrko B, Magasinski A, Hertzberg B, Milicev Z, Burtovyy R, Luzinov I, Yushin G (2011) Science 334:75
53. Feng J, Xiong S, Qian Y, Yin L (2014) Electrochim Acta 129:107
54. Zhang L, Zhang L, Chai L, Xue P, Hao W, Zheng H (2014) J Mater Chem A 2:19036
55. Yoon J, Oh DX, Jo C, Lee J, Hwang DS (2014) Phys Chem Chem Phys 16:25628
56. Liu J, Zhang Q, Wu Z-Y, Wu J-H, Li J-T, Huang L, Sun S-G (2014) Chem Commun 50:6386
57. Kumar PR, Kollu P, Santhosh C, Eswara Varaprasada Rao K, Kim DK, Grace AN (2014) New J Chem 38:3654
58. Li J, Zhao Y, Wang N, Ding Y, Guan L (2012) J Mater Chem 22:13002
59. Ryou M-H, Hong S, Winter M, Lee H, Choi JW (2013) J Mater Chem A 1:15224
60. Veluri PS, Mitra S (2013) RSC Adv 3:15132
61. Gao H, Zhou W, Jang J-H, Goodenough JB (2016) Adv Energy Mater 6 doi:10.1002/aenm.201502130
62. Chen C, Lee SH, Cho M, Kim J, Lee Y (2016) ACS Appl Mater Interfaces 8:2658
63. Chen Y, Liu N, Shao H, Wang W, Gao M, Li C, Zhang H, Wang A, Huang Y (2015) J Mater Chem A 3:15235
64. Chai L, Qu Q, Zhang L, Shen M, Zhang L, Zheng H (2013) Electrochim Acta 105:378
65. Tang H, Weng Q, Tang Z (2015) Electrochim Acta 151:27
66. Jeong YK, Kwon T-W, Lee I, Kim T-S, Coskun A, Choi JW (2014) Nano Lett 14:864
67. Wang J, Yao Z, Monroe CW, Yang J, Nuli Y (2013) Adv Func Mater 23:1194
68. Kwon T-W, Jeong YK, Deniz E, AlQaradawi SY, Choi JW, Coskun A (2015) ACS Nano 9:11317
69. Yoon D-E, Hwang C, Kang N-R, Lee U, Ahn D, Kim J-Y, Song H-K (2016) ACS Appl Mater Interfaces 8:4042
70. Murase M, Yabuuchi N, Han Z-J, Son J-Y, Cui Y-T, Oji H, Komaba S (2012) Chemsuschem 5:2307

71. Hwang G, Kim J-M, Hong D, Kim C-K, Choi N-S, Lee S-Y, Park S (2016) Green Chem 18:2710
72. Lee H, Yanilmaz M, Toprakci O, Fu K, Zhang X (2014) Energy Environ Sci 7:3857
73. Arora P, Zhang Z (2004) Chem Rev 104:4419
74. Kuribayashi I (1996) J Power Sour 63:87
75. Zhang LC, Sun X, Hu Z, Yuan CC, Chen CH (2012) J Power Sour 204:149
76. (a) Kim J-H, Kim J-H, Choi E-S, Yu HK, Kim JH, Wu Q, Chun S-J, Lee S-Y, Lee S-Y (2012) J Power Sour 242:533; (b) Chun S-J, Choi E-S, Lee E-H, Kim JH, Lee S-Y, Lee S-Y (2012) J Mater Chem 22:16618
77. Chen W, Shi L, Wang Z, Zhu J, Yang H, Mao X, Chi M, Sun L, Yuan S (2016) Carbohydr Polym 147:517
78. Leijonmarck S, Cornell A, Lindbergh G, Wagberg LJ (2013) Mater Chem A 1:4671
79. Pan R, Cheung O, Wang Z, Tammela P, Huo J, Lindh J, Edstroem K, Stroemme M, Nyholm L (2016) J Power Sour 321:185
80. Yu B-C, Park K, Jang J-H, Goodenough JB (2016) ACS Energy Lett 1:633
81. Jiang F, Yin L, Yu Q, Zhong C, Zhang J (2015) J Power Sour 279:21
82. Kelley J, Simonsen J, Ding J (2013) J Appl Polym Sci 127:487
83. Bolloli M, Antonelli C, Molmeret Y, Alloin F, Iojoiu C, Sanchez J-Y (2016) Electrochim Acta 214:38
84. Deng Y, Song X, Ma Z, Zhang X, Shu D, Nan J (2016) Electrochim Acta 212:416
85. Xu Q, Kong Q, Liu Z, Wang X, Liu R, Zhang J, Yue L, Duan Y, Cui G (2014) ACS Sustain Chem Eng 2:194
86. Xu Q, Kong Q, Liu Z, Zhang J, Wang X, Liu R, Yue L, Cui G (2014) RSC Adv 4:7845
87. Liao H, Zhang H, Hong H, Li Z, Qin G, Zhu H, Lin Y (2016) J Membr Sci 514:332
88. Liu C, Shao Z, Wang J, Lu C, Wang Z (2016) RSC Adv 6:97912
89. Long J, Wang X, Zhang H, Hu J, Wang Y (2016) Int J Electrochem Sci 11:6552
90. Yvonne T, Zhang C, Zhang C, Omollo E, Ncube S (2014) Cellulose (Dordrecht, Neth.) 21:2811
91. Zhang J, Yue L, Kong Q, Liu Z, Zhou X, Zhang C, Xu Q, Zhang B, Ding G, Qin B, Duan Y, Wang Q, Yao J, Cui G, Chen L (2014) Sci Rep 4:3935
92. Zhang J, Liu Z, Kong Q, Zhang C, Pang S, Yue L, Wang X, Yao J, Cui G (2013) ACS Appl Mater Interfaces 5:128
93. Alcoutlabi M, Lee H, Zhang X (2015) MRS Online Proc Libr 1718:1
94. Weng B, Xu F, Alcoutlabi M, Mao Y, Lozano K (2015) Cellulose (Dordrecht, Neth.) 22:1311
95. Huang F, Xu Y, Peng B, Su Y, Jiang F, Hsieh Y-L, Wei Q (2015) ACS Sustain Chem Eng 3:932
96. Lalia BS, Abdul Samad Y, Hashaikeh R (2012) J Appl Polym Sci 126:E441
97. Huang X (2014) J Power Sour 256:96
98. Yue Z, McEwen IJ, Cowie JMG (2003) Solid State Ionics 156:155
99. Regiani AM, De Oliveira Machado G, LeNest J-F, Gandini A, Pawlicka A (2001) Macromol Symp 175:45
100. Machado GO, Ferreira HCA, Pawlicka A (2005) Electrochim Acta 50:3827
101. Ledwon P, Andrade JR, Lapkowski M, Pawlicka A (2015) Electrochim Acta 159:227
102. Khanmirzaei MH, Ramesh S, Ramesh K (2015) Sci Rep 5:18056
103. Sato T, Banno K, Maruo T, Nozu R (2005) J Power Sour 152:264
104. Ren Z, Liu Y, Sun K, Zhou X, Zhang N (1888) Electrochim Acta 2009:54
105. Vasei M, Tajabadi F, Jabbari A, Taghavinia N (2015) Appl Phys A Mater Sci Process 120:869
106. Lin S-Y, Chen Y-C, Wang C-M, Wen C-Y, Shih T-Y (2012) Solid State Ionics 212:81
107. Neo CY, Ouyang JJ (2013) Mater Chem A 1:14392
108. Paracha RN, Ray S, Easteal AJ (2012) J Mater Sci 47:3698
109. Liu J, Li W, Zuo X, Liu S, Li Z (2013) J Power Sour 226:101
110. Zhao M, Zuo X, Wang C, Xiao X, Liu J, Nan J (2016) Ionics 22:2123

111. Abidin SZZ, Yahya MZA, Hassan OH, Ali AMM (2014) Ionics 20:1671
112. Chinnam PR, Zhang H, Wunder SL (2015) Electrochim Acta 170:191
113. Lee JM, Nguyen DQ, Lee SB, Kim H, Ahn BS, Lee H, Kim HS (2010) J Appl Polym Sci 115:32
114. Kang W, Ma X, Zhao H, Ju J, Zhao Y, Yan J, Cheng B (2016) J Solid State Electrochem 20:2791
115. Ngamaroonchote A, Chotsuwan C (2016) J Appl Electrochem 46:575
116. Mantravadi R, Chinnam PR, Dikin DA, Wunder SL (2016) ACS Appl Mater Interfaces 8:13426
117. Li M, Wang X, Wang Y, Chen B, Wu Y, Holze R (2015) RSC Adv 5:52382
118. Willgert M, Leijonmarck S, Lindbergh G, Malmstroem E, Johansson MJ (2014) Mater Chem A 2:13556
119. Chiappone A, Nair JR, Gerbaldi C, Bongiovanni R, Zeno E (2013) Cellulose (Dordrecht, Neth.) 20:2439
120. (a) Chiappone A, Nair JR, Gerbaldi C, Jabbour L, Bongiovanni R, Zeno E, Beneventi D, Penazzi N (2011) J Power Sour 196:10280; (b) Song M-K, Kim Y-T, Cho J-Y, Cho BW, Popov BN, Rhee H-W (2004) J Power Sour 125:10; (c) Chiappone A, Nair JR, Gerbaldi C, Bongiovanni R, Zeno E (2015). Electrochim Acta 153:97; (d) Chiappone A, Nair JR, Gerbaldi C, Zeno E, Bongiovanni R (2014) Eur Polym J 57:22; (e) Jafirin S, Ahmad I, Ahmad A (2013) BioResources 8:5947
121. Nair JR, Gerbaldi C, Chiappone A, Zeno E, Bongiovanni R, Bodoardo S, Penazzi N (2009) Electrochem Commun 11:1796
122. Kiristi M, Bozduman F, Gulec A, Teke E, Oksuz L, Oksuz AU, Deligoz HJ (2014) Macromol Sci Part A: Pure Appl Chem 51:481
123. Zhu YS, Xiao SY, Li MX, Chang Z, Wang FX, Gao J, Wu YP (2015) J Power Sour 288:368
124. Stojadinovic J, Dushina A, Trocoli R, La Mantia F (2014) ChemPlusChem 79:1507
125. Lu Y, Armentrout AA, Li J, Tekinalp HL, Nanda J, Ozcan SJ (2015) Mater Chem A 3:13350
126. Bella F, Nair JR, Gerbaldi C (2013) RSC Adv 3:15993
127. Yang Y, Hu H, Zhou C-H, Xu S, Sebo B, Zhao X-Z (2011) J Power Sour 196:2410
128. Liu K, Liu M, Cheng J, Dong S, Wang C, Wang Q, Zhou X, Sun H, Chen X, Cui G (2016) Electrochim Acta 215:261
129. Samad YA, Asghar A, Hashaikeh R (2013) Renew Energy 56:90
130. Salvador GP, Pugliese D, Bella F, Chiappone A, Sacco A, Bianco S, Quaglio M (2014) Electrochim Acta 146:44
131. Li P, Zhang Y, Fa W, Zhang Y, Huang B (2011) Carbohydr Polym 86:1216
132. Navarra MA, Dal Bosco C, Moreno JS, Vitucci FM, Paolone A, Panero S (2015) Membranes (Basel, Switz.) 5:810
133. Jin L, Greene GW, MacFarlane DR, Pringle JM (2016) ACS Energy Lett 1:654
134. (a) Yamazaki S, Takegawa A, Kaneko Y, Kadokawa J-I, Yamagata M, Ishikawa M (2009) Electrochem Commun 11:68; (b) Yamazaki S, Takegawa A, Kaneko Y, Kadokawa J-I, Yamagata M, Ishikawa M (2010) J. Power Sour 195:6245; (c) Yamazaki S, Takegawa A, Kaneko Y, Kadokawa J-I, Yamagata M, Ishikawa M (2010) J Electrochem Soc 157:A203
135. Yamagata M, Soeda K, Ikebe S, Yamazaki S, Ishikawa M (2012) ECS Trans 41:25
136. Stephan AM, Kumar TP, Kulandainathan MA, Lakshmi NA (1963) J Phys Chem B 2009:113
137. Osman Z, Ibrahim ZA, Arof AK (2000) Carbohydr Polym 44:167
138. Andrade JR, Raphael E, Pawlicka A (2009) Electrochim Acta 54:6479
139. Wang S, Zhang Q-X, Xu Y-Z, Li D-M, Luo Y-H, Meng Q-B (2013) J Power Sour 224:152
140. Ma G, Yang Q, Sun K, Peng H, Ran F, Zhao X, Lei Z (2015) Bioresour Technol 197:137
141. Hou J, Cao C, Ma X, Idrees F, Xu B, Hao X, Lin W (2014) Sci Rep 4:7260
142. Jain A, Xu C, Jayaraman S, Balasubramanian R, Lee JY, Srinivasan MP (2015) Microporous Mesoporous Mater 218:55
143. Song S, Ma F, Wu G, Ma D, Geng W, Wan JJ (2015) Mater Chem A 3:18154
144. Chen H, Liu D, Shen Z, Bao B, Zhao S, Wu L (2015) Electrochim Acta 180:241

145. Wang J, Shen L, Xu Y, Dou H, Zhang X (2015) New J Chem 39:9497
146. Li Y, Zhang Q, Zhang J, Jin L, Zhao X, Xu T (2015) Sci Rep 5:14155
147. Li Y, Yu N, Yan P, Li Y, Zhou X, Chen S, Wang G, Wei T, Fan Z (2015) J Power Sour 300:309
148. Wang P, Wang Q, Zhang G, Jiao H, Deng X, Liu L (2015) J Solid State Electrochem. Ahead of Print
149. Ma G, Ran F, Peng H, Sun K, Zhang Z, Yang Q, Lei Z (2015) RSC Adv 5:83129
150. Fan Z, Qi D, Xiao Y, Yan J, Wei T (2013) Mater Lett 101:29
151. Peng C, Lang J, Xu S, Wang X (2014) RSC Adv 4:54662
152. Lv Y, Gan L, Liu M, Xiong W, Xu Z, Zhu D, Wright DS (2012) J Power Sour 209:152
153. Li X, Xing W, Zhuo S, Zhou J, Li F, Qiao S-Z, Lu G-Q (2011) Biores Technol 102:1118
154. Rufford TE, Hulicova-Jurcakova D, Zhu Z, Lu GQ (2008) Electrochem Commun 10:1594
155. Liew SY, Thielemans W, Freunberger S, Spirk S (2017) Polysaccharide based supercapacitors. Springer, Berlin
156. Kim KW, Kim JS, Lee SW, Lee JK (2015) Electrochim Acta 170:146
157. Huang Y, Zheng M, Lin Z, Zhao B, Zhang S, Yang J, Zhu C, Zhang H, Sun D, Shi YJ (2015) Mater Chem A 3:10910
158. Wang L, Schutz C, Salazar-Alvarez G, Titirici M-M (2014) RSC Adv 4:17549
159. Huang Y, Lin Z, Zheng M, Wang T, Yang J, Yuan F, Lu X, Liu L, Sun D (2016) J Power Sour 307:649
160. Chen W, Deng D (2015) Carbon 87:70
161. Cao S, Feng X, Song Y, Liu H, Miao M, Fang J, Shi L (1073) ACS Appl Mater Interfaces 2016:8
162. Li Y, Zhu H, Shen F, Wan J, Han X, Dai J, Dai H, Hu L (2014) Adv Funct Mater 24:7366
163. Zhu H, Shen F, Luo W, Zhu S, Zhao M, Natarajan B, Dai J, Zhou L, Ji X, Yassar RS, Li T, Hu L (2017) Nano Energy 33:37
164. Shen F, Zhu H, Luo W, Wan J, Zhou L, Dai J, Zhao B, Han X, Fu K, Hu L (2015) ACS Appl Mater Interfaces 7:23291
165. Zhu H, Jia Z, Chen Y, Weadock N, Wan J, Vaaland O, Han X, Li T, Hu L (2013) Nano Lett 13:3093

Chapter 3
Conclusion and Outlook

Although many publications exist which report on polysaccharides as components in batteries, particularly as binder and separators, their full potential has not been fully exploited yet. This can be attributed that mainly cellulose derivatives have been used and systematic approaches to improve the device performance by specifically altering the polysaccharide backbone have not been demonstrated so far. As a consequence, there are innumerous options to further improve different type of battery components to improve the performance of batteries. However, the largest potential can be seen in the development of new SPE and GPE systems since intense research in those areas has just begun a few years ago, whereas cellulose based binders and separators have been employed for quite some time. All these efforts must take into account challenges which are not directly related to the chemistry but to the engineering and device fabrication point of view which have a significant impact on the overall device performance, rather than material performance on its own.

Acknowledgements The Austrian Research Promotion agency for funding by the Research Studios Austria for Advanced and Innovative Materials for Electrochemical Energy Storage (Grant Number 844759) and the COMET K-project FLIPPR[2].

Printed in the United States
By Bookmasters